Dissections

Plane & Fancy

Can you cut an octagon into five pieces and rearrange them into a square? How about turning a star into a pentagon? These are just two of the many challenges of geometric dissections, the mathematical art of cutting figures into pieces that can be rearranged to form other figures, using as few pieces as possible. This book shows you many ingenious ways to solve these problems and the beautiful constructions you can create.

Through the ages, geometric dissections have fascinated puzzle fans and great mathematicians alike. Here you will find dissections known to Plato alongside exciting new discoveries. The author poses puzzles for you to solve, but this is much more than a puzzle book. He explains solution methods carefully: new and old types of slides, strips, steps, tessellations, and exploration of star and polygon structures. You need only a basic knowledge of high school geometry.

You will also be introduced to the people – famous, not-so-famous, and obscure – who have worked on these problems. Travel from the palace school of tenth-century Baghdad to the mathematical puzzle columns in turn-of-the-century newspapers, from the 1900 Paris Congress of Mathematicians to the night sky over Canberra.

This beautifully illustrated book will provide many hours of enjoyment for any mathematical puzzle enthusiast.

Dissections

Plane & Fancy

GREG N. FREDERICKSON
Purdue University

CAMBRIDGE
UNIVERSITY PRESS

PUBLISHED BY THE PRESS SYNDICATE OF THE UNIVERSITY OF CAMBRIDGE
The Pitt Building, Trumpington Street, Cambridge, United Kingdom

CAMBRIDGE UNIVERSITY PRESS
The Edinburgh Building, Cambridge CB2 2RU, UK
40 West 20th Street, New York, NY 10011-4211, USA
10 Stamford Road, Oakleigh, Melbourne 3166, Australia
Ruiz de Alarcón 13, 28014 Madrid, Spain
Dock House, The Waterfront, Cape Town 8001, South Africa

http://www.cambridge.org

First published 1997
First paperback edition 2002

Printed in the United States of America

Typeface Lucida Bright

A catalog record for this book is available from the British Library

Library of Congress Cataloging in Publication data is available

ISBN 0 521 57197 9 hardback
ISBN 0 521 52582 9 paperback

To Susanne, who sees my smiles,
to Nora, who hears my music, and
to Paul, who solves my puzzles.

Contents

	Preface	*page* ix
1	"Dat Pussle"	1
2	Our Geometric Universe	9
3	Fearful Symmetry	20
4	It's Hip to Be a Square	28
5	Triangles and Friends	40
6	All Polygons Created Equal	51
7	First Steps	60
8	Step Right Up!	71
9	Watch Your Step!	89
10	Just Tessellating	105
11	Plain Out-Stripped	117
12	Strips Teased	136
13	Tessellations Completed	150
14	Maltese Crosses	157
15	Curves Ahead	163
16	Stardom	172

17	Farewell, My Lindgren	187
18	The New Breed	207
19	When Polygons Aren't Regular	221
20	On to Solids	230
21	Cubes Rationalized	247
22	Prisms Reformed	258
23	Cheated, Bamboozled, and Hornswoggled	268
24	Solutions to All Our Problems	278
	Bibliography	293
	Index of Dissections	303
	General Index	307

Preface

As befits a book on recreational mathematics, this one has been great fun to write, in part because I let my curiosity lead the way. It was a revelation to survey the original puzzle columns of Dudeney in *The Strand Magazine* of the 1910s and 1920s. It was fascinating to dig out even earlier references and make connections that other authors had missed. It was intriguing to collect biographical information on the people who have made a contribution to this area.

I hope that the book will also be great fun to read. The intended audience is anyone who has had a course in high school geometry and thought that regular hexagons were rather pretty. I have used some other high school math here and there. If you find some of the formulas tough sledding, you should be able to skip over them without much lost. Likewise, if you skip over the algorithmlike descriptions of methods in Chapters 7–9, you won't lose much but can be comforted that precise descriptions do exist. On the other hand, if you get intrigued with some topic and want to follow up on it, I have provided additional comments and references, ordered by chapter, in the Afterword.

It has been a surprise to see how many people have contributed to whatever success this book may enjoy. First are those who have produced new dissections, in quality and quantity substantially greater than I had imagined. I gratefully acknowledge the permission of Duilio Carpitella, Anton Hanegraaf, Bernard Lemaire, David Paterson, Robert Reid, Gavin Theobald, and Alfred Varsady to reproduce their unpublished or privately published dissections.

Martin Gardner is the unofficial godfather of this book. A letter from him got me thinking about dissections again in 1991. And when I started on the book in 1994, he was generous with his help and advice. Some of his leads were crucial. He forwarded a decade-old letter from Robert Reid that led to an especially fruitful interaction. And Martin also put me in touch with Will Shortz, and that contact led to a fortunate chain of events. Will generously provided me with the citations to the earlier versions of dissection puzzles that later appeared in Sam Loyd's

Cyclopedia. And Will also told Jerry Slocum what I was up to. Jerry clued me in to the trade cards that Sam Loyd had produced, and he shared copies from his puzzle collection with me. Jerry also alerted me to David Singmaster's wonderful work on sources of recreational mathematics. David shared an electronic copy of his book, and xerox copies of Dudeney's puzzle columns in the *Weekly Dispatch*, too. David's sources book helped me focus on earlier historical material. It also led to my contact with Anton Hanegraaf, who generously shared his treasure trove of unpublished material with me. Anton also provided a scholarly critique of a draft of my first chapter on solid dissections, accompanied by an additional annotated bibliography.

Friends new and old have helped me improve the text with corrections and suggestions: Reva Chandrasekaran, Jaydeep Chipalkatti, David Eppstein, Bill and Gisela Fitzgerald, Susanne Hambrusch, Ravi Janardan, Edgar Knapp, Robert Reid, Susan Rodger, Jerry Slocum, and John Woodrum.

Others have helped in a variety of technical ways: Erika Gautschi, Concettina Guerra, and Susanne Hambrusch helped with languages other than English. Sam Wagstaff identified number theory references. Bryn Dole helped reproduce the Sam Loyd puzzle graphics. Adam Hammer and Dan Trinkle helped me navigate through the software jungle on the computer system.

Thanks to those who helped track down people who have created dissections: Martin Gardner, H. Martin Cundy, Joseph Madachy, and Harry Nelson.

I would like to express my appreciation to the following people, who have generously given assistance in locating or supplying biographical information: Carol Arrowsmith, Peter Cadwell, Diana Chardin, Mary Collison, Dirk Ferus, William A. Freese, Richard Funkhouser, Beat Glaus, Jeremy R. Goldberg, Elisabeth Hambrusch, Susanne Hambrusch, Tony Heathcote, Martina Leitsch, Judy Lindgren, Chip Martel, Frank R. Miles, Mrs. J. M. Morris, Dorothy E. Mosakowski, Robert A. Rosenbaum, Jane Sadler, Doris Schneider-Wagenbichler, B. J. Stokes, Kim Walters, Heather Whitby, Michael Yeates. I would especially like to thank Judy Lindgren, who surveyed a mountain of her late father's correspondence for me.

I would like to thank the following people and institutions that made reference materials available to me: Apollonia Lang Steele, the library reference staff at Purdue University, and the Interlibrary Loan staff not only at Purdue University and but also at all the libraries that lent materials. I would like to acknowledge the following libraries that I visited: Brown University libraries, Tippecanoe County (Indiana) Public Library, Library of Congress, the library of the Smithsonian Institution, Milwaukee Public Library, the library of the Chicago Historical Society, and the Newberry Library in Chicago.

I gratefully acknowledge the permission of the following publishers to reproduce copyrighted material in this book: Figure 10.9 is reproduced from *Recreational Mathematics Magazine* with the kind permission of Joseph S. Madachy. Figure 14.12 and the figure in Solution 14.6 are reproduced from *Le Monde* with their kind permission. Figures 9.4, 11.13, and 12.11 are reproduced from the *American Mathematical Monthly* with the kind permission of the Mathematical Association of America. Figures 11.16, 11.17, 11.23, and 12.15-17.17 are reproduced from the *Mathematical Gazette* with the kind permission of the Mathematical Association. Figures 1.1, 16.10, 16.17, 17.6, 17.8, and 17.12 are reproduced from *Recreational Problems in Geometric Dissections, and How to Solve Them,* by Harry Lindgren, 1972, with the kind permission of Dover Publications, Inc. Figures 10.5, 11.19, 11.22, 12.3, 12.14, 12.17, 12.18, 13.3, and 23.11-23.14 are reproduced from the *Australian Mathematics Teacher* with the kind permission of the Australian Association of Mathematics Teachers, Inc. Figures 1.2, 6.17, 9.6, 9.8, 9.18-9.20, 10.25, 11.25, 11.26, 16.16, 16.20, 17.7, 17.13, 17.15, 17.18, 17.19, 17.24, 17.25, 17.27, 18.13, 18.17, 22.1, 22.2, 22.5-22.12, and the figure in Solution 9.2 are reproduced from the *Journal of Recreational Mathematics,* ©1973, 1974, 1978, 1979, 1982, 1983, 1985, 1986, 1989 Baywood Publishing Company, Inc.

I would like to thank Ian Stewart for giving his encouragement and for providing the carrot and stick that kept me going near the end. I would also like to thank Lauren Cowles, at Cambridge University Press, whose suggestions helped me improve the presentation in the book. Finally, I would like to thank Rena Wells, whose care and professionalism made book production a surprisingly enjoyable task.

In preparing the paperback edition, I have limited myself to correcting known errors, rather than introducing dissections discovered since the original hardcover edition was published five years ago. There have been enough new discoveries so that major portions of the text would have to be typeset once again — a tedious and expensive task given the need to place figures and biographies carefully. Instead, I refer readers to the webpages that I have set up:

http://www.cs.purdue.edu/homes/gnf/book.html

Surely there will be more exciting discoveries in the future, so that the webpages are the preferred way to sustain the book as a living document. Also, I have posted photos of many dissection friends and have listed events, reviews, and other information related to the book. I hope that readers will enjoy this additional material.

CHAPTER 1

"Dat Pussle"

A small hand reaches up, and an eager smile appears: "I wanna-pay-wit dat pussle." From the drawer indicated, the brown puzzle pieces are retrieved, then laid flat upon the carpet. The not-yet-three-year-old sets about his task with the words "I make a diamond." The "diamond" is a regular heptagon, and the set makes a pair of heptagons, so the dad helps his son divide up the pieces and get started. With some coaching, the son forms first one, and then the other, of the heptagons. Finally, the dad takes his turn, assembling the whole set into a 7-pointed star, with a little help from his son. Repeated many times already, this activity has transcended puzzle into ritual – a ritual that echoes the dad's creation of the puzzle two decades earlier.

The puzzle in the preceding reminiscence was a physical model of the dissection shown in Figure 1.1. The two heptagons on the right are divided by line segments into nine pieces, which then rearrange to form the 7-pointed star on the left. A geometric dissection is a cutting of one or more figures into pieces that can be rearranged to form other figures. Since the constructions are precise, this activity falls within the realm of mathematics, though it is of a decidedly recreational flavor. This is a book devoted solely to such mathematical constructions.

A dissection can be appreciated in several ways. The most basic is as a puzzle in which a given set of pieces must be assembled to form a desired figure. Grown-ups as well as two-year-olds can enjoy this activity. Many adults have been intrigued by the task of assembling the three-dimensional star-prisms in Figure 22.8.

It is also fun to provide a puzzle to a friend and supply timely hints for its solution. And the pleasure is heightened if you construct the model yourself. If

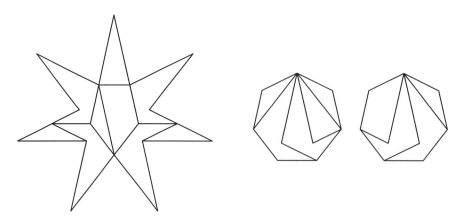

1.1: Seven-pointed star and two heptagons

you enjoy woodworking and understand some high school geometry, you can build many of the puzzles in this book.

However, you need not make a trip to the hardware store or the art supply center to enjoy this book. You can sit back in a comfy armchair, leaf through the book, and revel in the figures as graphic art or mathematical objects. The dissection of 6-pointed stars in Figure 5.24 is a nice example.

The main goal in a dissection puzzle is to find one that has as few pieces as possible and yet as much symmetry as possible. Many of the chapters contain puzzles, which range from easy to hard, with solutions appearing at the end of the book. You can also enjoy dissections by discovering them yourself, by pushing beyond the boundaries of the book. The dad in the story started with Lindgren's book (1964b), then posed and solved new dissection puzzles, and also improved solutions to existing puzzles.

There is a surprisingly rich history associated with dissections. I have attempted to trace each dissection as far back as possible and have identified some notable events. For example, dissecting two equal squares to one larger square in four pieces dates back at least to Plato. In the tenth century, several beautiful dissections were described by Arabian mathematicians in their commentaries on Euclid's *Elements*. During the Renaissance, the Italian physician and mathematician Girolamo Cardano described a simple, basic dissection in his compendium of scientific knowledge.

Dissections even played a role in the determination of the value of π. A regular polygon of n sides approximates a circle when n is large. Thus the area of the polygon is a good approximation for the area of the circle, πr^2. When $n = 12$,

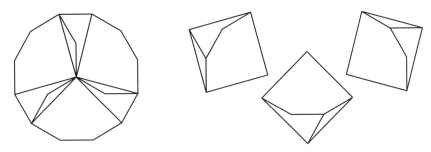

1.2: Variation of Tai Chen's dodecagon to three squares

we get the rough approximation of 3 for π, since the area of the dodecagon exactly equals $3r^2$, or 3 times the area of a square whose side length is r. An elegant dissection proof of this fact appears in one of the illustrations that the eighteenth-century Chinese scholar Tai Chen prepared for the palace edition of the *Chiu Chang Suan Shu* (Nine Chapters on the Mathematical Art). The text of the manuscript is from the third-century commentary of Liu Hui. Unfortunately, earlier diagrams by Liu Hui were lost centuries ago. More recently Kürschák (1899) gave a similar proof. Elliott (1986) has given a symmetrical variation of Tai Chen's dissection (Figure 1.2).

The eighteenth and nineteenth centuries produced a variety of puzzle books that contained dissection puzzles. Noteworthy examples are those by Panckoucke (1749), Ozanam (1778), Jackson (1821), Lucas (1883), Lemon (1890), and Hoffmann (1893). Given the importance of geometry in both the academic curriculum and the industrial world of the nineteenth century, it is not surprising that articles on dissections appeared in journals such as *Journal für die reine und angewandte Mathematik, Transactions of the Royal Society of Edinburgh, Journal Asiatique, Messenger of Mathematics,* and *Nouvelle Correspondance Mathématique.*

In the early nineteenth century, Lowry (1814), Wallace (1831), Bolyai (1832), and Gerwien (1833) showed how to dissect any set of two-dimensional figures bounded by straight lines into a finite number of pieces that can be rearranged to form another such set of equal area. This result has commonly been referred to as the Bolyai–Gerwien theorem, as in (Boltyanskii 1963). In three dimensions, the analogous property does not hold. Hilbert (1900) included as the third in his famous list of unsolved problems the problem of showing that there exist two polyhedra of equal volume such that one cannot be dissected into a finite number of pieces that rearrange to form the other. Within a few months, Dehn (1900) had proved Hilbert's conjecture, though variations of the problem continued to be of interest for decades afterward, as discussed by Boltyanskii (1978).

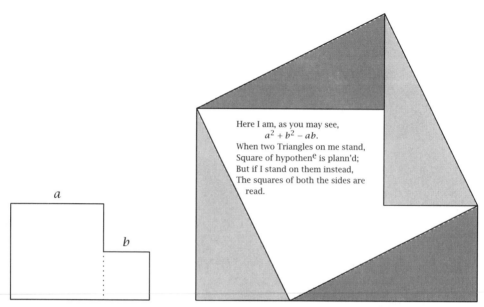

Here I am, as you may see,
$a^2 + b^2 - ab$.
When two Triangles on me stand,
Square of hypothen^e is plann'd;
But if I stand on them instead,
The squares of both the sides are read.

1.3: Two squares　　　　　**1.4:** George Biddle Airy's poem

By virtue of its beauty and simplicity, one dissection was the catalyst for poetic sentiment in the mid-nineteenth century. An 1855 letter from Augustus DeMorgan to W. Rowan Hamilton, collected in (Graves 1889), reproduced an announcement by George Biddle Airy of a simple dissection proof of the Pythagorean theorem. Airy included a poem that refers to squares of side lengths a and b, attached along one side as shown (reduced in size) in Figure 1.3. The dissection plus poem are in Figure 1.4, with the white piece and the lower two right triangles forming the outline of the figure in Figure 1.3. The white piece plus the upper two right triangles form a single, tilted square. Airy wrote the poem from the viewpoint of the white piece, which is of area $a^2 + b^2 - ab$. The term ab represents the combined area of two right triangles with legs of length a and b. Airy, the astronomer royal at Greenwich from 1836 to 1881, pointed out that the two right triangles could be moved via translation, with no rotation, from one configuration to the other. The insight behind the dissection itself seems to predate Airy, as I discuss at the beginning of Chapter 4.

In the latter half of the nineteenth century, the use of manufacturers' trade cards for advertising became all-pervasive. As described by Burdick (1967) and Jay (1987), these cards attained their greatest popularity during the 1880s and 1890s, when they were collected by the American public. Some companies featured puzzles on their trade cards. Slocum (1996) reproduces a selection of eight puzzle cards from the period 1873–1892, of which five contain a dissection puzzle. Of the eight cards, Jerry Slocum attributes four to the American puzzlist Sam Loyd.

Near the close of the nineteenth century, puzzle columns by Sam Loyd and the English puzzlist Henry E. Dudeney began to appear in newspapers and magazines. These two collaborated early on in the English periodical *Tit-Bits*, with Loyd initially writing the puzzles and Dudeney, under the pseudonym "Sphinx," writing the commentary and awarding prize money. Ambitious and competitive, these men found their relationship strained by a natural rivalry.

Loyd and Dudeney soon went their separate ways and presided over a remarkable upsurge of interest in mathematical puzzles on both sides of the Atlantic. In due course their puzzles were collected into anthologies that retain their charm today: (Loyd 1914), (Gardner 1959), (Gardner 1960), (Dudeney 1907), (Dudeney 1917), (Dudeney 1926a), (Dudeney 1931a), (Dudeney 1931b), (Dudeney 1967). Loyd and Dudeney were skilled not only in designing engaging puzzles, but also in clothing them in elaborate presentations. Loyd was perhaps the more impressive, as we see from his Sedan Chair Puzzle:

> ... These chairs are made of rattan wicker work and remind you very much of those little Chinese puzzle boxes, made of colored straws so cleverly put together that you cannot discover where they are joined together – all of which is suggestive of a clever puzzle, for those sedan chairs will close up so as to make a covered box when it rains, and yet the closest examination will not detect where the pieces are joined.... You are asked to cut the sedan into the fewest possible pieces which will fit together and form a perfect square.

The solution to the Sedan Chair Puzzle uses a neat quarter-turn and requires just two pieces (Figure 1.5). Each line segment that is neither horizontal nor vertical is the hypotenuse of a right triangle with legs of 3 and 4 units in length. I have indicated the bases of these triangles with dotted lines in the square. With the base of the sedan chair equal to 11 units, the resulting square will be 11 units on a side.

The twentieth century brought an increase in sophistication to dissections. The later puzzle columns of Dudeney broke new ground, posing puzzles that were sufficiently challenging that their published solutions were often bettered, sometimes dramatically. Articles and problems appeared in periodicals such as the *Mathematical Gazette, American Mathematical Monthly, Australian Mathematics Teacher, Scientific American*, and *Journal of Recreational Mathematics*. (This last journal seems to have become the preferred outlet for articles on geometric dissections.)

THE SEDAN CHAIR PUZZLE.

PROPOSITION—Show how to close the sedan chair.

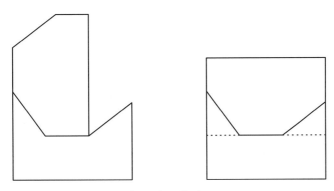

1.5: Loyd's Sedan Chair to square

Geometric dissection has become a frequent topic in books on mathematical recreations. See, for example, books by Fourrey (1907), Ball (1939), Kraitchik (1942), Mott-Smith (1946), Gardner (1961), Domoryad (1964), Cadwell (1966), Gardner (1966), Gardner (1969), Kordemsky (1972), van Delft and Botermans (1978), and Gardner (1986). There is a chapter on dissection theory in the two-volume geometry textbook by Eves (1963). Gardner (1956) includes a chapter on dissection paradoxes, and numerous entries on dissections appear in the compilation of interesting geometrical facts by Wells (1991). Singmaster (1995) gives a detailed historical review of a number of dissection puzzles and paradoxes. Dissections figure prominently in a text produced by the Education Development Center (1997) as part of a curriculum revision with the goal of making mathematics more palatable to high school students. Finally, books have been written that focus exclusively on dissections: (Lindgren 1964b), its revision (Lindgren 1972), (Boltyanskii 1963), and (Boltyanskii 1978).

My original plan for this book was less ambitious. However, a check of some sources uncovered surprises, which encouraged a more thorough historical review. I decided to group material by dissection type and to open each chapter by evoking the context of the earliest dissections of that type.

As I searched back to the sources, my curiosity about the people who contributed to this esoteric subject grew. Were they butchers, bakers, candlestick makers? Were they mathematical amateurs or pros? Did their day-to-day activities require a certain degree of technical ability? As I tracked down the famous, the not-so-famous, and the obscure, my curiosity grew to fascination, and my fascination to near-obsession. I have encapsulated the results of this search in short biographies that are distributed throughout the book. I hope that you will enjoy making the acquaintance of these individuals. And one pattern was confirmed: Cambridge

University had by far the most contributors. No other institution comes close – check it out in the biographies!

In collecting historical and biographical information, I consulted some relatively inaccessible sources, especially in tracking down the earliest appearance of many dissections. If you are interested in following up some of the references, try referring first to some of the more accessible books, such as (Lindgren 1972), (Gardner 1959), (Gardner 1960), (Dudeney 1907), (Dudeney 1917), (Dudeney 1926a), (Dudeney 1931a), (Dudeney 1931b), and (Dudeney 1967).

One phenomenon that I did not fully anticipate was that as I organized and analyzed previous dissections, new ones would suggest themselves in profusion. And these new dissections have provided the nicest of surprises and the most pleasant of diversions. Some personal favorites are the variety of hinged dissections in Chapter 3 and elsewhere; two squares to a different pair of squares to yet a different pair of squares in Chapter 8; two hexagrams to one in Figure 5.24; the special-case dissections of two stars to one in Figures 18.22 and 18.23; the conversions of the P-slide and Q-slide techniques to step techniques in Figures 7.2, 9.5, and 9.7; the step dissections for classes of squares in Chapters 7 and 8; the polyhedral tessellation dissections of Chapter 10; and seven heptagons to one, and eight heptagons to one, in Figures 17.30 and 17.31. I hope that the excitement that has accompanied these discoveries shines through in their presentation.

Another pleasant surprise has been the continuing creativity of contemporary dissection enthusiasts. Especially memorable have been the days when packets containing record-breaking dissections have arrived in the mail from around the world: from David Paterson, Alfred Varsady, Robert Reid, Gavin Theobald, and Anton Hanegraaf. The treasure troves of dissections by Reid, by Theobald, and by Hanegraaf were completely unanticipated and could well have remained buried, had it not been for serendipitous circumstances. (Are there others whose work has not yet come to light?) At the hands of these individuals, I have suffered the same fate as earlier puzzlists, including Sam Loyd, Henry Dudeney, and Harry Lindgren, in the sense that some of my dissections have been improved upon. However, this competition to find the best solution is what drives the field forward. So it is particularly appropriate to give this activity of "going one better," as Harry Lindgren called it, special prominence throughout the book.

CHAPTER 2

Our Geometric Universe

One case after another, the Renaissance scholar identified the ways in which the regular polygons can tile the plane. Four centuries later, his thoroughness might be viewed as routine. And yet his diagrams belie this view – those marvelous diagrams! One after another, yes, but with ever-increasing intricacy, with regular figures regularly arranged, few-sided with many-sided polygons, some with stars.

Johannes Kepler was investigating what he called the congruence *of polygons. Those select few polygons that possessed this property formed the basis of his* Harmony of the Universe, *the remarkable amalgam of geometry, music, and astronomy by which he sought to explain the motions of the heavens. This work contained his masterpiece, the third law of planetary motion. It also asserted that the six (known) planets revolved around the sun in orbits isolated by the shells of the five Platonic solids! It was mysticism, devout religious belief, and great science all rolled up into one. It was fabulous.*

Johannes Kepler was evidently the first person to perform a unified study of how polygons can fit together to tile the plane. It appeared in Book II of Part V of *Harmonice Mundi*, collected in (Kepler 1940). It is impossible to know whether Kepler would have performed this study outside the framework of *Harmonice Mundi*. But the tilings that he investigated are so beguiling that he probably could not have resisted. These tilings of the plane, or tessellations, will be essential to understanding many of the dissections of plane figures.

In this chapter we make a bold tour of our geometric universe. First we will visit galaxies of plane figures, including polygons and stars, and examine their structure.

We start with polygons. A *polygon* is a closed plane figure bounded by straight lines. The simplest type of polygon is a *regular polygon*, in which all sides are of equal length and all angles are equal. We shall use the notation $\{p\}$ to represent a regular polygon with p sides and focus on those regular polygons for which $p \leq 12$:

triangle: $\{3\}$	hexagon: $\{6\}$	enneagon: $\{9\}$
square: $\{4\}$	heptagon: $\{7\}$	decagon: $\{10\}$
pentagon: $\{5\}$	octagon: $\{8\}$	dodecagon: $\{12\}$

At times we shall also dissect polygons that are not regular, representing a p-sided polygon that is not regular by $\{\bar{p}\}$. For example, $\{\bar{4}\}$ is an unrestricted 4-sided polygon, or *quadrilateral*.

Some irregular triangles and squares are "more regular" than others. An *isosceles triangle* has two sides that are equal. The two angles that are opposite the equal sides are also equal. A *trapezoid* is a quadrilateral with two opposite sides parallel to each other; a *parallelogram* is a quadrilateral with each side parallel to the side opposite it; and a *rhombus* is a parallelogram with all four sides of equal length. The two angles adjacent to any edge in a rhombus sum to 180°, or π radians. We shall refer to a rhombus with angles of 72° and 108° as a 72°-rhombus, or a $2\pi/5$-rhombus.

We can think of a regular polygon as a union of various rhombuses (or rhombi) and isosceles triangles (which are halves of rhombuses). Such a characterization is an *internal structure* of the corresponding polygon. If the regular polygon has an even number of sides, then the internal structure needs only rhombuses. We see representative internal structures for $\{4\}$, $\{6\}$, $\{8\}$, $\{10\}$, and $\{12\}$ in Figure 2.1. A regular polygon with $2q$ sides contains q each of $(i\pi/q)$-rhombuses, for each

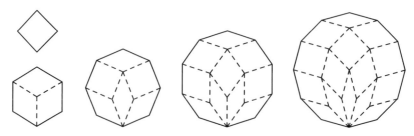

2.1: Internal structures for $\{4\}$, $\{6\}$, $\{8\}$, $\{10\}$, and $\{12\}$

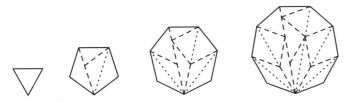

2.2: Internal structures for {3}, {5}, {7}, and {9}

positive integer i that is less than $q/2$, plus $q/2$ squares if q is even. For example, the dodecagon contains 6 each of $(\pi/6)$-rhombuses and $(\pi/3)$-rhombuses, plus 3 squares.

If the regular polygon has an odd number of sides, then the internal structure contains some isosceles triangles. We see representative internal structures for {3}, {5}, {7}, and {9} in Figure 2.2. The structure for the pentagon was known to the ancient Greeks; the principles underlying the structure of the heptagon and enneagon were known to Arabian mathematicians of the late tenth century. Our treatment follows an unpublished description by Alfred Varsady. A regular polygon with $2q + 1$ sides contains $2i - 1$ isosceles triangles whose third angle equals $(2i-1)\pi/(2q+1)$, for $i = 1, \ldots, q$. All but one of each type of isosceles triangles are paired to give a rhombus, with the paired triangles separated by a dotted line, and the rhombuses and remaining triangles separated by dashed lines.

If the regular polygon has a number of sides that is a multiple of 3, then there are alternative internal structures based on equilateral triangles (Figure 2.3).

We venture deeper into our universe to study *star* polygons. We form the boundary of a regular polygon {q} by placing q vertices at equal intervals around the circumference of a circle and connecting each vertex to its clockwise neighbor with a straight line segment. If $q > 4$, then we can generate a star, denoted by {$q/2$}, by connecting each vertex not to its neighbor but to the second nearest vertex in a clockwise direction. Thus we construct a {5/2} as on the left of Figure 2.4. Take the

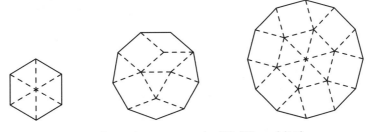

2.3: Alternative structures for {6}, {9}, and {12}

2.4: Forming {5/2} and {9/3}

star to be the area enclosed within the line segments, and the boundary to be those portions of the line segments that separate the star from the rest of the plane. The {5/2}, also called a *pentagram*, is shown in the middle of the figure. In general, for $1 < p < q/2$, we have a star {q/p}, where we connect each vertex to its pth nearest vertex in a clockwise direction. Thus we have started to form a {9/3} on the right of Figure 2.4.

Our definition of stars is more generous than that in sources such as (Eves 1963), where p and q must be relatively prime. Using that definition would forfeit not only the 6-pointed star but a galaxy of other beautiful stars. For that reason, we follow in the tradition of dissection experts such as Sam Loyd and Harry Lindgren.

We can assign an internal structure to stars, in much the same way as for regular polygons. For any star {q/p} for which q is even, we can restrict the structure to rhombuses only. Examples are shown for {6/2}, {8/2}, {8/3}, {10/2}, {10/3}, and {10/4} in Figure 2.5. The description of a structure for {8/2} contains a structure for {8/3}, and a similar observation holds for the 10-pointed stars.

There is a fairly simple structure for {5/2}, but structures for odd-pointed stars are in general more complicated than for even-pointed stars. However, an

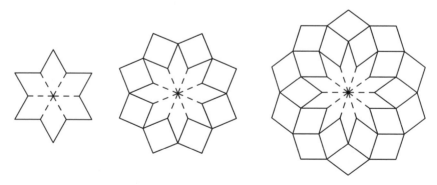

2.5: Structures for {6/2}, {8/2}, {8/3}, {10/2}, {10/3}, and {10/4}

2.6: Internal structures for {5/2}, {9/2}, {9/3}, and {9/4}

odd-pointed star whose number of points is a multiple of 3 is more cooperative. Each of the {9/2} and {9/4} in Figure 2.6 contains a {9/3}. We can produce the {9/3} by attaching nine equilateral triangles to an enneagon, which we have already seen in Figure 2.3. The structure for {9/4} contains rhombuses and half-rhombuses whose side lengths are smaller than the side length of one of the points of the {9/4}. This is not such an unusual occurrence. We shall see alternative internal structures for many polygons and stars for which the side lengths of the rhombuses do not equal the side lengths of the polygon or star.

Besides the regular polygons and stars, we find other remarkable objects lurking in our universe. Perhaps the most noteworthy are the crosses. The simplest crosses are those consisting of identical squares attached together. The Greek Cross, denoted by {G}, consists of five squares; the Latin Cross, denoted by {L}, consists of six squares; and the Cross of Lorraine, denoted by {L'}, consists of thirteen squares (Figure 2.7).

2.7: Greek Cross, Latin Cross, Cross of Lorraine

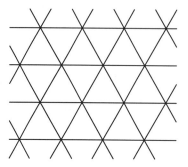

2.8: Tessellation: triangles

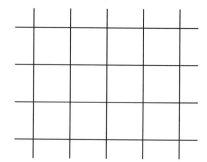

2.9: Tessellation: squares

We will explore the Maltese Cross and some of its transmogrifications in Chapter 14 and curved figures such as crescents, oval stool tops, and the like in Chapter 15.

Now that we have made a brief survey of the planar figures in our geometric universe, let's see how they can tile the plane, in the spirit of Johannes Kepler. Given a set of planar figures, a *tessellation* is a covering of the plane with copies of figures from the set such that the figures do not overlap. The figure that is used to tile that plane is a *tessellation element* and may consist of one or more pieces.

A *regular tessellation* is a tiling with just one type of regular polygon, such that a vertex of a polygon touches a side of another polygon only at an endpoint of that side (which is a vertex of that polygon). There are only three regular tessellations, of triangles, of squares, and of hexagons, in Figures 2.8, 2.9, and 2.10.

A *semiregular tessellation* allows more than one type of regular polygon, but again a vertex may touch a side only at an endpoint of that side, and the same sequence of polygons must appear in order around each vertex. Including the regular tessellations, there are eleven semiregular tessellations, of which we show the

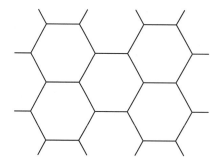

2.10: Tessellation: hexagons

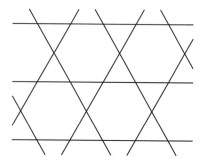

2.11: Hexagons and triangles

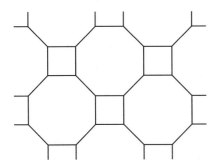

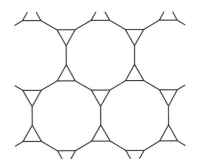

2.12: Octagons and squares   **2.13:** Dodecagons and triangles

three most useful for geometric dissections: hexagons and triangles in Figure 2.11, octagons and squares in Figure 2.12, and dodecagons and triangles in Figure 2.13.

There are also tessellations of the plane in which a vertex of one polygon is coincident with an interior point of an edge of another polygon. These tessellations thus allow two adjacent polygons to meet at edges that do not completely coincide; we call them *non-edge-to-edge* tessellations. Two examples, involving triangles and squares, respectively, are shown in Figures 2.14 and 2.15, respectively.

Non-edge-to-edge tessellations also exist for triangles and hexagons together, where the side lengths of the triangles are different from the side lengths of the hexagons. As shown in Figures 2.16 and 2.17, these come in two flavors, depending on which polygon has the larger side length. These tessellations result in a natural way from perturbing the tessellation in Figure 2.11.

If we allow polygons to have different edge lengths, then we can match small squares with large squares, as in Figure 2.18. We see in Figure 2.19 that the same arrangement works for similar rectangles.

These latter two figures have tiny circles to indicate what we call the *repetition pattern* of the tessellation. Each circle identifies a point in the tessellation, and these

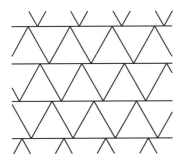

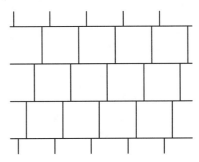

2.14: Non-edge-to-edge triangles   **2.15:** Non-edge-to-edge squares

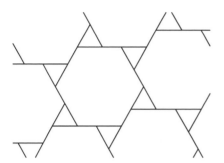

2.16: Big hexagons, small triangles

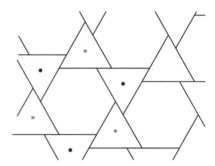

2.17: Big triangles, small hexagons

points are identical in their positioning within the polygon or set of polygons used to form the tessellation. We can connect the points to form a convex figure of area equal to that of the set of polygons used to form the tessellation. We can then tile the plane with this convex figure. In the case of Figure 2.18, the repetition pattern is that of a square. There is nothing special about the set of points chosen; we could just as well have translated the points by any fixed distance in any fixed direction. In the case of Figure 2.19, the repetition pattern is that of another rectangle that is larger than, but similar to, the rectangles in the tessellation.

There are other repetition patterns besides the ones for squares and rectangles. In the case of Figures 2.16 and 2.17, we can position the points to be the centers of triangles, giving a repetition pattern that is a hexagon. Since three rather than six hexagons meet at a vertex in a tessellation of hexagons, we partition the set of repetition points into two sets. In Figure 2.17, for example, one set (tiny circles) consists of the centers of triangles pointing down, and the other set (tiny asterisks) consists of centers of triangles pointing up.

Finally, let's look at two tessellations involving nonconvex figures. In Figure 2.20 we give a tessellation of Greek Crosses. The repetition pattern of this tessellation is

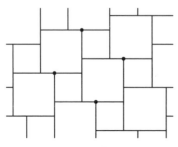

2.18: Unequal squares

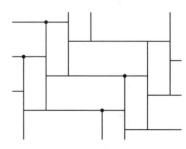

2.19: Unequal rectangles

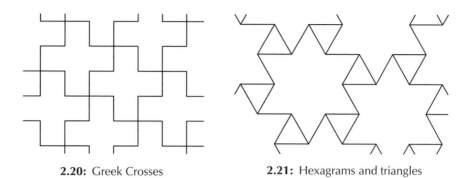

2.20: Greek Crosses   **2.21:** Hexagrams and triangles

a square. In Figure 2.21 we give a tessellation of {6/2}s and twice as many triangles. The repetition pattern of this tessellation is a hexagon.

Having explored a planar swath of our universe, we now look for the three-dimensional objects in "deep space." The ancient Greeks discovered the three-dimensional analogues of regular polygons, namely, regular polyhedra. A *regular polyhedron* is a three-dimensional figure formed by taking one type of regular polygon for its faces and joining the polygons at the vertices so that the same number of polygons fit together at any vertex. We shall use the notation $\{p, q\}$ to denote a regular polyhedron with q faces of type $\{p\}$ meeting at each vertex. Triangles give rise to three different polyhedra, two of which are shown in Figure 2.22: The *tetrahedron*, $\{3, 3\}$, has four triangular faces, with three meeting at any vertex. The *octahedron*, $\{3, 4\}$, has eight triangular faces, with four meeting at any vertex. The *icosahedron*, $\{3, 5\}$, has twenty triangular faces, with five meeting at any vertex. The square and the pentagon each give rise to one polyhedron each, shown in Figure 2.23: The *cube*, or *hexahedron*, $\{4, 3\}$, has six square faces, with three meeting at any vertex. The *dodecahedron*, $\{5, 3\}$, has twelve pentagonal faces, with three meeting at any vertex.

Another set of three-dimensional figures are the *Archimedean solids*, catalogued by the ancient Greek Archimedes. There are thirteen Archimedean solids, and these,

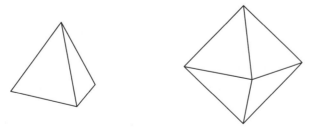

2.22: Tetrahedron and octahedron

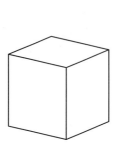
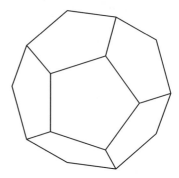

2.23: Cube and dodecahedron

plus two infinite families, the prisms and the antiprisms, constitute the set of semiregular polyhedra that are not regular. A *semiregular polyhedron* is a three-dimensional figure formed by taking regular polygons for its faces, with the polygons joined at the vertices so that the same sequence of polygons appears in order around each vertex. We illustrate one Archimedean solid, the truncated octahedron, in Figure 2.24. Denoted by $t\{3, 4\}$, it has six square and eight hexagonal faces, with two hexagons and one square meeting at each vertex. It is the only semiregular polyhedron, besides the cube, that by itself can tile (or *honeycomb*) three-dimensional space.

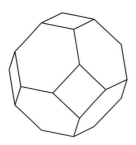
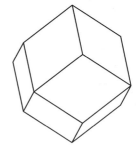

2.24: Truncated octahedron   **2.25:** Rhombic dodecahedron

We can form some three-dimensional figures by joining identical copies of a given irregular polygon together. In particular, twelve rhombuses enclose the *rhombic dodecahedron*, where the diagonals of each rhombus are in the ratio of $\sqrt{2} : 1$. Illustrated in Figure 2.25, it can easily be formed from two equal cubes, cutting one of the cubes into six equal pyramids whose apexes are in the center, and then attaching one pyramid to each face in the other cube. As suggested by this construction, the rhombic dodecahedron can also honeycomb three-dimensional space.

A final class of three-dimensional figures that we dissect are prisms. A *prism* is bounded on the *top* and *bottom* by two copies of a planar figure oriented in the same way in two parallel planes. The *sides* of the prism are parallelograms, with each parallelogram containing as two of its boundary edges a corresponding pair of edges from the copies of the planar figure. When all resulting sides are perpendicular to the top and bottom, the sides are rectangles and the prism is a *right prism*. We shall focus on right prisms, and use the term *prism* to mean a right prism, unless otherwise stated. A triangular prism is shown in Figure 2.26, and an {8/2}-prism is shown in Figure 2.27.

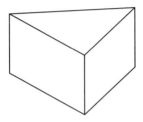

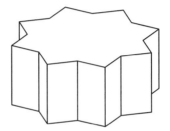

2.26: Triangular prism   **2.27:** {8/2}-prism

CHAPTER 3

Fearful Symmetry

When the stars threw down their spears
And water'd heaven with their tears:
Did he smile his work to see?
Did he who made the Lamb make thee?

Tyger, Tyger, burning bright,
In the forests of the night:
What Immortal hand or eye,
Dare frame thy fearful symmetry?

Dare we attempt to frame a fearful metaphor, using the last two stanzas of William Blake's "The Tyger," from his *Songs of Experience* (1794)? Might the structure of polygons and stars be a sort of geometric meter, and the arrangement of cuts in the dissections be geometric rhyme? Do not our stars and other dissected figures luminesce in a cold mathematical stillness? Do not the best of the dissections fit together with a cunning that is diabolical? Dare we illustrate the preceding excerpt with a haloed rendition of Harry Lindgren's dissection of a {12/5} to a Latin Cross (Figure 3.1)?

As the rhyme in Blake's poem is not absolute, so also is the symmetry in Lindgren's dissection not absolute. Lindgren's dissection comes close but does not possess reflection symmetry. *Reflection symmetry* is present when there exists a line of reference such that what is on one side of the line is exactly mirrored, or reflected, on the other side. As shown in Figure 3.2, if we split just one of the pieces into two pieces, then both the star and the cross will possess reflection symmetry about the dotted lines. The moral is that by relaxing symmetry, we sometimes get a more economical dissection. The mystery is that this is so unpredictable.

3.1: Dissection of {12/5} to {L}: Tyger and Lamb?

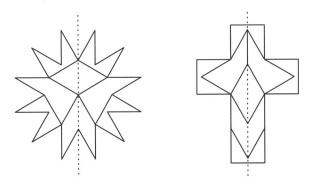

3.2: Symmetrical but nonminimal {12/5} to {L}

Another form of symmetry is *rotational symmetry*, in which a rotation about the center of a figure produces a figure that coincides with the original figure. The dissection of two equal Greek Crosses to a square from (Loyd, *Tit-Bits*, 1897a) is an example. The dissected square appears identically oriented after a quarter-turn, so that it possesses 4-fold rotational symmetry. Similarly, each Greek Cross in the dissection possesses 2-fold rotational symmetry.

The dissection of two equal Greek Crosses to a square also possesses two other notable types of symmetry. The first is translation with no rotation. The second is

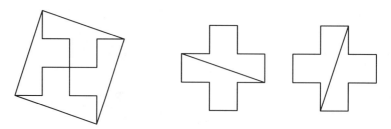

3.3: Loyd's two Greek Crosses to a square

replication symmetry. In *replication symmetry*, a particular dissected figure appears more than once in the dissection. In Figure 3.3, the Greek Crosses are cut in precisely the same manner, so that they exhibit 2-fold replication symmetry.

A dissection possesses *translation with no rotation*, or is *translational*, if we can move the pieces from one figure to the other without rotating them. It is not always possible to find dissections between two figures that possess this property, no matter how the figures are oriented with respect to each other. Hadwiger and Glur (1951) gave a simple characterization of which figures can have a dissection that has this property. For each figure, walk around its boundary in a clockwise direction and identify the slopes of each line segment in the boundary. Each line segment has a slope and a direction, either to the right or to the left, over which it is walked. (For vertical line segments, replace right and left by up and down.) For each figure, and for each slope of line segments in that figure, add the lengths of the line segments walked to the right and subtract from this sum the lengths of line segments walked to the left. If for every slope, the sum for one figure equals the sum for the other figure, then and only then is there a translational dissection.

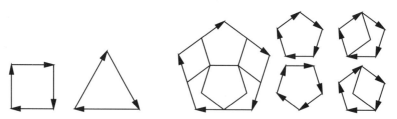

3.4: Square versus triangle   **3.5:** Four pentagons to one

An example is the pair consisting of the square and triangle in Figure 3.4. There is no translational dissection of a triangle to a square, because the sums of opposite pairs of sides in the square sum to zero, whereas each of the three sides of the triangle contributes a nonzero value that is not offset by any other side. The dissection by Harry Lindgren (1964b) of four pentagons to one, shown in Figure 3.5, is an example that is translational. With one of the small pentagons oriented the opposite of the other three, the sums of side lengths for each of five different slopes are the same for the four small pentagons as for the large one. However, even when a pair of figures satisfies Hadwiger and Glur's requirement, it may not be possible to find a dissection that both uses the minimum number of pieces and is translational. An example is the dissection of the Latin Cross and {12/5} in Figure 3.1. The sum for each slope in the Latin Cross is zero, and similarly for the {12/5}, so that there is a translational dissection. However, it seems unlikely that there is any such dissection that has as few as seven pieces.

Lindgren's dissection of four pentagons to one also possesses other types of symmetry. The large pentagon exhibits reflection symmetry, and, when suitably arranged, so does the set of small pentagons. The small pentagons also possess what we shall call partial replication. *Partial replication* is present when only a proper subset of identical figures are identically cut. In this case, two of the small pentagons are identically cut, and the other two are uncut (and thus identically cut!).

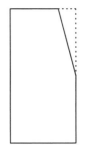
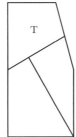

3.6: Clipped rectangle   **3.7:** Loyd's 3-piece dissection

Although a dissection may be more attractive if no pieces are turned over, sometimes achieving a minimum dissection seems to require that at least one piece be turned over. An example is the "clipped rectangle" in Figure 3.6. The original rectangle is twice as high as it is wide, and the triangle is clipped off it with a cut from the midpoint of the right side at an angle of 15° from the vertical. We have indicated the removed triangle with dotted lines. Loyd (*Tit-Bits*, 1897c) posed the problem of dissecting such a clipped rectangle to a square in three pieces.

The dissection is shown in Figure 3.7, with the piece that is turned over marked with a T in the clipped rectangle, and with a circled T in the resulting square. One cut starts at the midpoint of the slanted side and is of length equal to the width of the clipped rectangle. In the comments accompanying the puzzle, "Sphinx" stated that the angle need not be restricted to 15°. Lindgren (1970) showed that the solution technique will work for angles up to 22.5°, but only if the dimensions of the original rectangle are also changed. If the angle is α, then the height of the rectangle is $(1 + 2\sin(2\alpha))$ times its width, and the clip begins at a point above the base at a height equal to the rectangle's width.

Puzzle 3.1: Draw the 3-piece dissection of a clipped rectangle to a square when the angle is exactly 22.5°.

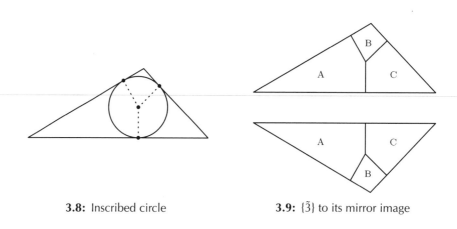

3.8: Inscribed circle   **3.9:** {3̄} to its mirror image

We can use a simple trick when turning over is not allowed. To dissect an irregular triangle to its mirror image, first inscribe a circle in the triangle, as shown in Figure 3.8. Dots indicate the center of the inscribed circle and the points of tangency with the triangle's three sides. Dotted lines, which connect the center to these three points, correspond to cuts in the top triangle of Figure 3.9. Each of the three resulting pieces has reflection symmetry. Thus none needs to be turned over to achieve the mirror image of the original triangle. Boltyanskii (1978) gives this two-dimensional analogue of a dissection, which we will see in Figures 20.2 and 20.3. Any irregular polygon that can have a circle inscribed in it can be converted to its mirror image by this technique. The number of pieces will equal the number of vertices in the polygon.

A dissection is *rational* if each cut is parallel to some line segment in the boundary and is of a length that is a rational number times the length of that line segment. Loyd (*Tit-Bits*, 1897b) presented a dissection of a square to two Greek Crosses, where one cross was one-quarter the area of the other. All the cuts in Figure 3.10 are either

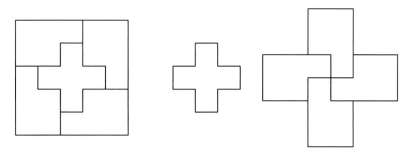

3.10: Rational dissection of a square to two Greek Crosses

horizontal or vertical and are one or two times the side length of the smaller Greek Cross.

A dissection is *fully hinged* if the pieces are connected with hinges into a chain so that rotating the pieces around one way assembles them into one of the figures of the dissection, and rotating the pieces the other way assembles them into the other figure of the dissection. The fully hinged dissection in Figure 3.11 consists of three pieces connected by two hinges. In one orientation, the pieces form a large square, whereas in the other, they make a figure whose outline consists of a small

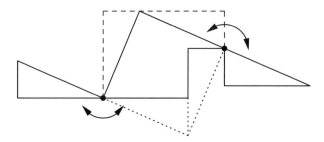

3.11: Fully hinged: two squares to one

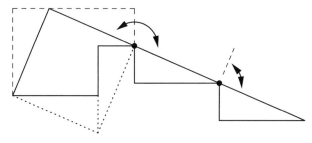

3.12: Variously hinged: two squares to one

square flush against a medium-sized square. The hinges are designated by small circles. (A physical model might use metal hinges if the pieces are wood, or a sturdy tape if the pieces are cardboard.) An unhinged version of this dissection is shown in Figure 4.4.

An intriguing feature of this dissection is that we can fully hinge it in two other ways. One of these is shown in Figure 3.12. A dissection is *variously hinged* if we can hinge it with the largest possible number of hinges in several different ways.

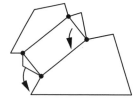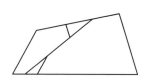

3.13: Cyclicly hinged: one quadrilateral to another

When we can remove a hinge without disconnecting a hinged set of pieces, we shall call the dissection *cyclicly hinged*. An example in Figure 3.13 is a cyclicly hinged version of Harry Lindgren's dissection of one quadrilateral to a second quadrilateral with the same angles but different base and height. We can swing the top piece either clockwise or counterclockwise to close the parallelogram-shaped gap between the pieces. We will learn more about this dissection when we get to Figure 5.4.

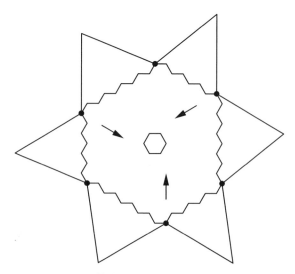

3.14: Partially hinged: two hexagons to one

For some dissections we can hinge most, rather than all, of the pieces. We shall call such a dissection *partially hinged*. As an example, we see a hinged version of hexagons of side lengths 1 and $\sqrt{48}$ to a hexagon of side length 7 in Figure 3.14. Six of the pieces are cyclicly hinged, but since the seventh piece is isolated, the dissection is also partially hinged. When we move three of the hinges simultaneously in the directions shown, we form the hexagon of side 7. When we move the other three hinges simultaneously toward the center and remove the small hexagon, then we form the hexagon of side $\sqrt{48}$. Try forming the hexagons from these pieces, and then check your solution in Figure 9.26.

CHAPTER 4

It's Hip
to Be a Square

The year is 1701. You have just entered the Rose and Crown, a bookshop in St. Paul's Churchyard, London, England. The bookseller, recalling your interest in books scientific and mathematical, pulls out a recent translation of a 1695 book by a German academic, Johann Christoph Sturm: Mathesis Enumerata: or, the Elements of the Mathematicks. *"This, esteemed Sir, is a book most handsome. I beseech you: Inspect the many fine figures." At his behest, you leaf through the thin volume, examining the plates. One of the figures catches your eye: a novel illustration of the* Pythagorick Theorem. *It places side by side two squares, then cuts them into five pieces that rearrange to form a larger square. Indeed, it is elegant enough to prompt you to review the mathematicks of your purse.*

The dissection that caught your eighteenth-century eye is shown in Figure 4.2. It gives a physical realization of a proof of the Pythagorean theorem, which is Proposition 47 of Book I of Euclid's *Elements of Geometry*. Sturm (1700) produced the illustration, based on a proof by the seventeenth-century Dutch mathematician Frans van Schooten. However, not mentioned in the book is the fact that the dissection is actually much older, having made an appearance eight hundred years before, in Thābit ibn Qurra's *Risāla fi'l-hujja al-mansūba ilā Suqrāt fi'l-murabba wa qutrihi* (Treatise on the Proof Attributed to Socrates on the Square and Its Diagonals).

al-Sābi al-Harrānī Thābit ibn Qurra was was born in Harran, Mesopotamia (now Turkey), in 836. A moneychanger in his youth, Thābit was invited to Baghdad by the mathematician Mūsā ibn Shākir because of his knowledge of Syriac and Greek. There he learned mathematics and astronomy, and eventually he became a great scholar. He translated Euclid's *Elements*, Ptolemy's *Almagest*, and works of Archimedes. He wrote a large number of treatises on algebra, geometry, and trigonometry, as well as many on astronomy. His work laid the groundwork for the definition of real numbers, the eventual discovery of integral calculus, and the eventual development of non-Euclidean geometry. He also wrote on mechanics, much on medicine, and on philosophy and humanistic sciences. In his last years, Thābit was a member of the caliph's retinue. He died in Baghdad in 901.

Thābit's is one of two general 5-piece dissections of two squares to one. We can derive both dissections by using a technique in which we superpose tessellations, although we have no evidence that either was originally discovered that way. Also, in this chapter we dissect a pair of squares to a different pair of squares, and three squares to a large square. For the latter puzzle, we examine a second general technique, called the P-slide.

The technique of *superposing tessellations* creates two tessellations with the same repetition pattern and overlays them so that the repetition points coincide. Once the tessellations are superposed, the line segments in one tessellation induce cuts in the figures of the other.

If we superpose the tessellation of the two squares from Figure 2.18 with a tessellation of the large square, as shown in Figure 4.1, we get the first dissection, in Figure 4.2. Paul Mahlo (1908) realized that he could produce this dissection by superposing tessellations, as did Percy MacMahon (1922). Brodie (1884) observed that a dissection like the one in Figure 4.2 also applies to similar rectangles. This gives a 5-piece dissection whenever the long side of the smaller rectangle is shorter

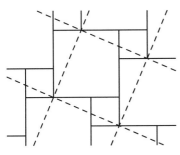

4.1: Mahlo's superposition

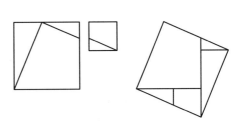

4.2: Two squares to one

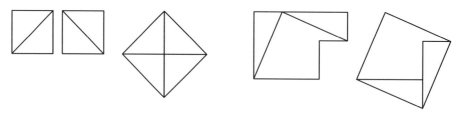

4.3: Two equal squares   **4.4:** Two attached squares

than the short side of the larger rectangle, and a 4-piece one when those sides are equal. Superposing tessellations gives this dissection too. These dissections are all translational.

In the case of two equal squares, one piece disappears, giving the 4-piece dissection in Figure 4.3. This was given in Plato's *Menon* and also in his *Timaeus*. If the two given squares in Figure 4.2 are attached, so as to make one piece, then the dissection simplifies to the 3-piece dissection in Figure 4.4. In fact, the pieces in Sturm's version of Figure 4.2 were given three different shades, suggesting the three pieces in Figure 4.4.

The second dissection of two squares to one, by Henry Perigal (1873), results from shifting the placement of one tessellation relative to the other, as shown in Figure 4.5. This dissection, shown in Figure 4.6, is translational and can be partially hinged. The resulting large square has 4-fold rotational symmetry.

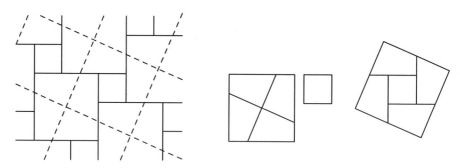

4.5: Tessellations for Perigal's dissection   **4.6:** Perigal's dissection

Suppose we superpose two tessellations of the type in Figure 2.18, as shown in Figure 4.7? We have then found a way to dissect a pair of squares to a different pair of squares. The resulting 6-piece dissection (Figure 4.8) is a natural generalization of Perigal's dissection. The larger square in each pair has 4-fold rotational symmetry and translation with no rotation. When each pair of squares is attached, so that we

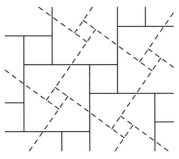

4.7: Tessellations generalized

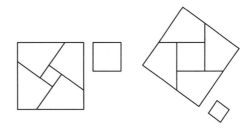

4.8: Two squares to two

are dissecting one figure to one other figure, the dissection simplifies to four pieces, which can be fully and variously hinged.

A third, more challenging puzzle is to dissect three squares to a single square. The simplest version is for three squares of equal side length. Abū'l-Wafā, a tenth-century mathematician and astronomer, gave a pretty 9-piece dissection (Figure 4.10), in which the large square has 4-fold rotational symmetry. David Wells (1975) found the following derivation: Convert two of the squares to a larger one, using the

Henry Perigal was a London stockbroker who lived from 1801 to 1899. An amateur mathematician, he had his dissection of two squares to one printed on his business card. The dissection was also carved on the south front face of his monument in the churchyard of Wennington, in Essex. Perigal was a lifelong bachelor, who spent his evenings at a variety of scientific meetings: He was treasurer of the Royal Meteorological Society for almost fifty years and was also a member of the London Mathematical Society, the Royal Microscopic Society, the Royal Institution, and some fifteen other societies and clubs. After attending meetings of the Royal Institution as a visitor for many years, he became a member on his *94th birthday*.

In astronomical matters, Henry Perigal was what was termed a "paradoxer": His primary belief was that the moon did not rotate. In a heroic effort to convert others, he drew diagrams, built models, and wrote poems. Despite these views, and reflecting the charm of his personality, he was elected first a fellow of the Royal Astronomical Society and later a member of that group's inner circle. Perigal was also noted to be an excellent lathe worker and wrote on numerous subjects, including the geometry of lathe work, the construction methods of the pyramids, harmonic motion, and cycloidal curves. His nickname "Cyclops" was given him by a lady friend, who had mistaken it for *cycloids*, a word she had heard frequently mentioned in connection to him.

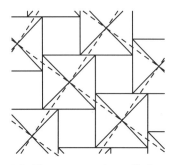

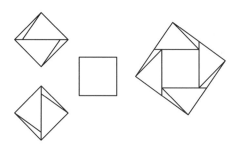

4.9: Three squares' tessellations

4.10: Abū'l-Wafā's dissection

dissection in Figure 4.3. Then use Perigal's dissection on the new and the remaining squares, superposing tessellations as in Figure 4.9. Perhaps we should attribute Perigal's dissection to Abū'l-Wafā.

Abū'l-Wafā's dissection is not minimal. Paul Busschop found a 7-piece dissection, which was reproduced in (Catalan 1873). This dissection was based on a strip technique (see Chapter 11). Henry Dudeney (*Dispatch*, 1900a) gave the 6-piece dissection in Figure 4.11. Two years before Dudeney was born, Philip Kelland (1855) published a 4-piece dissection of a "gnomon" (consisting of three attached squares) to one square, as shown in Figure 4.12. One of the cuts is along the edge of attachment of two of the squares, and adding another cut (shown as dashed) along the other edge of attachment gives precisely the dissection presented by Dudeney. This same dissection, but with the dashed edges actually included as dashed edges, was given by Perigal (1891).

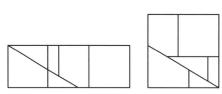

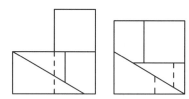

4.11: Perigal's three squares

4.12: Kelland's gnomon

The technique used in the dissections of Figures 4.11 and 4.12 is what Harry Lindgren calls a *parallelogram slide*, or *P-slide*. It converts one parallelogram to another with the same angles but a different base, using three pieces. The bases must be within a factor of 2 of each other. The P-slide is illustrated in Figure 4.13, where the parallelogram ABCD with the longer base is on the left. We locate point E on the base so that line segment BE is the length of the desired shorter base, and

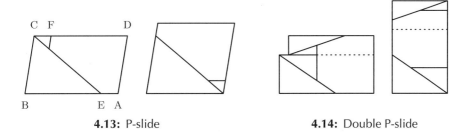

4.13: P-slide   **4.14:** Double P-slide

make cut CE. Similarly, we locate point F on the top edge so that BE and FD are of equal length. Then we cut down from F along a line parallel to side BC, stopping when it intersects the first cut. We slide the two pieces that are above CE past each other along this line to form the parallelogram on the right. Kelland's dissection is the earliest use of the P-slide that we know of.

Philip Kelland was a British mathematician who lived from 1808 to 1879. He distinguished himself as a mathematics student at Queen's College, Cambridge, earning an M.A. and also taking holy orders there. In 1838 he was appointed professor of mathematics at the University of Edinburgh, an exceptional appointment for an Englishman with a completely English education.

Kelland was active in the movement advocating reform of the Scottish university system. A founder of the Scottish Life Association, he made a tour in the United States and Canada as part of the septennial investigation in 1858. Occasionally he officiated at various Episcopal churches in Edinburgh.

Kelland published papers in both physics and mathematics, most notably on the motion of waves in canals, various problems in optics, non-Euclidean geometry, and algebra. He was president of the Royal Society of Edinburgh at the time of his death.

Figure 4.14 shows how to use P-slides in conjunction with each other. We view the given figure as two rectangles joined on an edge, so that we can apply a P-slide to each of these rectangles. The nontriangular pieces remain adjacent in each of the transformed rectangles, so that they remain joined along the dotted lines in the resulting 5-piece dissection. This *double P-slide* technique is shown in the dissection in Figure 4.18. Perhaps the earliest use of a double P-slide was by George Wotherspoon in his 10-piece dissection of the heptagon to the square, given in (Dudeney, *Strand*, 1927b).

When three squares are not all equal, dissecting them is more challenging. Let the side lengths of the three given squares be x, y, and z, and the side length of the resulting square be w. I identify three cases that cover all possibilities, with each case handled by an 8-piece dissection. David Paterson (1995a) discussed dissections of n

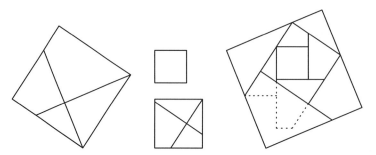

4.15: Uncentered version of Perigal's dissection used twice

unequal squares to m unequal squares briefly. For certain relative sizes of squares, he claimed $(4n - 3)$-piece dissections for $m = 1$, and $(4n + 4m)$-piece dissections for $m > 1$. I assume for the next three dissections that $x \leq y \leq z$.

My first 8-piece dissection is for the case when $z > (2x^2 + y^2)/\sqrt{x^2 + y^2}$. I start by applying an uncentered version of Perigal's dissection to the x-square and the y-square, and then an uncentered version again to the resulting square and the y-square. This gives the 9-piece dissection shown with solid lines in Figure 4.15. To save a piece, merge the four pieces of the y-square into just two. The merging is suggested by the dotted lines, which outline two pieces from the y-square that end up repositioned and then merged. To make space, a fifth piece must be cut from the z-square. Saving two pieces and creating one gives the 8-piece dissection in Figure 4.16.

When $z \leq (2x^2 + y^2)/\sqrt{x^2 + y^2}$, there are two related cases. In each one, use a P-slide to convert the y-square to a $(y^2/(w - x) \times (w - x))$-rectangle and the z-square to a $((w - x) \times z^2/(w - x))$-rectangle. Nestle the x-square in the lower right corner of the w-square, the first rectangle in the upper right, and the second rectangle in the lower left. When $y^2 > (w - x)x$, this results in the two rectangles

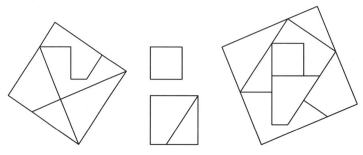

4.16: Squares when $z > (2x^2 + y^2)/\sqrt{x^2 + y^2}$

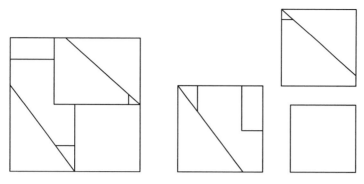

4.17: Three squares to one: P-slides and overlap

overlapping. However, we can cut the overlap from the second rectangle and place it in the upper left corner, as shown in Figure 4.17. When $y^2 < (w - x)x$, there is no overlap, but the second rectangle extends beyond the top boundary of the w-square. Then we cut off this extension and place it between the rectangles. This latter dissection works as long as $x + 2y \geq w$. When $x + 2y < w$, then $z > (2x^2 + y^2)/\sqrt{x^2 + y^2}$.

So far we have handled the cases of three equal squares and three unequal squares. Between these extremes are a wealth of special relationships among x, y, z, and w that allow for dissections with fewer pieces. Some of these relationships also make possible efficient rational dissections, which we explore in Chapter 8. I list the special relationships in the following, dropping the restriction $x \leq y \leq z$.

$$z^2 = x^2 + y^2 \tag{4.1}$$

$$z = y \tag{4.2}$$

$$w = x + y, \qquad \text{with } x < y \quad \text{and} \quad z < 4x \tag{4.3}$$

$$z = 2(w - y), \qquad \text{with } x > z \tag{4.4}$$

$$w + x = y + z, \qquad \text{with } x < y < z \tag{4.5}$$

$$z^2 = y^2 + (w - x)^2, \qquad \text{with } x < \frac{3w}{4} \tag{4.6}$$

Relationship (4.1) makes a nice puzzle:

Puzzle 4.1: Find a 7-piece dissection for squares realizing $x^2 + y^2 + z^2 = w^2$ in the case that $x^2 + y^2 = z^2$.

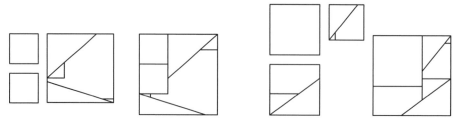

4.18: Three squares: two equal I   **4.19:** Three squares: two equal II

There are three subcases for (4.2). If $x > \sqrt{2}y$, then we have the 7-piece dissection in Figure 4.18. Just stack the two equal squares one on top of the other and cut the remaining square by using a double P-slide.

If $y/2 < x < \sqrt{2}y$, then we have the 7-piece dissection in Figure 4.19. The dissection results by performing a P-slide on the x-square, and a similar operation on the y-square. The trick is to locate the cuts for a P-slide on a $(y \times (w-x^2)/(w-y))$-rectangle, and then to remove the extra height to give a y-square.

Puzzle 4.2: Find a 5-piece dissection of three squares x, y, and z to another square, where $z = y$ and $x = \sqrt{2}y$.

For relationship (4.3), we have the 6-piece dissection shown in Figure 4.20. We place the x-square on top of the y-square, cut a $((y-x) \times y)$-rectangle out of the y-square, rotate it, and fill the rectangular cavity above the rectangle by cutting the remaining square with a P-slide.

A 7-piece dissection for (4.4) is shown in Figure 4.21. We cut the x-square out of the upper left corner of the w-square, and the z-square out of the upper right corner. Then a P-slide makes the base be of length w, and a second P-slide shifts the top of the remaining portion of the y-square.

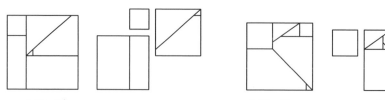

4.20: Three squares: $w = x + y$ **4.21:** Three squares: $z = 2(w - y)$

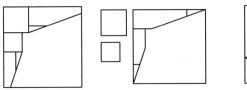

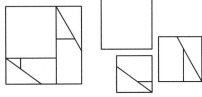

4.22: Three squares: filling in   **4.23:** Three squares: reshaping

We can use a variant of a double P-slide to produce a 7-piece dissection for (4.5) in Figure 4.22. We place the y-square in the upper left corner of the w-square, and the x-square beneath it. Imagine the remaining part of the w-square partitioned into three rectangles: an $(x \times w{-}x{-}y)$-rectangle below the x-square, a $(y \times w{-}y)$-rectangle to the right of the y-square, and a $(w{-}x \times w{-}y)$-rectangle. When the dimensions satisfy relationship (4.5), then we can change the first two rectangles via P-slides into rectangles that together with the $(w{-}x \times w{-}y)$-rectangle fill in the z-square.

Paul Busschop (1876) gave an 8-piece dissection for the special case, $(\sqrt{2})^2 + (\sqrt{3})^2 + 2^2 = 3^2$, of relationship (4.6). He applied the dissection of Figure 4.2 twice, first to convert the $\sqrt{2}$-square and 2-square to a $\sqrt{6}$-square, and then to convert the $\sqrt{3}$-square and resulting $\sqrt{6}$-square to a 3-square. Because of the same ratios on side lengths of the squares, one of the cuts in the second application of Figure 4.2 coincides with a cut from the first application. However, we use two P-slides to get a 7-piece dissection for all of (4.6), in Figure 4.23. We cut the x-square out of the upper left corner of the w-square. Then we partition the remaining part of the w-square into two rectangles: an $(x \times w{-}x)$-rectangle below the x-square, and a $(w{-}x \times w)$-rectangle. When the dimensions satisfy this relationship, we can reshape the first rectangle via a P-slide into the y-square, and the other via a P-slide into the z-square.

Puzzle 4.3: Find a 6-piece dissection for squares realizing the relation $2^2 + 3^2 + (\sqrt{3})^2 = 4^2$.

Two isolated instances are also interesting. First, Geoffrey Mott-Smith (1946) gave a 7-piece dissection for $1^2 + (\sqrt{3})^2 + (\sqrt{5})^2 = 3^2$. Second, I found a 7-piece dissection for $2^2 + (\sqrt{8})^2 + (\sqrt{13})^2 = 5^2$ (Figure 4.24). It uses the trick from Figure 4.4 plus an observation made by Loyd (*Eagle*, 1896a): A 3-square that has lost a 1-square from a corner has a 3-piece dissection to a square.

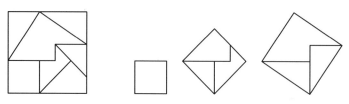

4.24: Squares: $\sqrt{13}$, $\sqrt{8}$, 2

Let's close the chapter with a dissection puzzle that is related to two squares to one. The law of cosines states that for any triangle with side lengths x, y, and z, and angle θ opposite side z, $z^2 = x^2 + y^2 - 2xy\cos\theta$. The last term is positive when $\theta > 90°$. In this case, a parallelogram with sides of length x and y and the smaller of its angles $\theta - 90°$ has area $-xy\cos\theta$. Rudolf Hunger (1921) posed and solved the puzzle of dissecting a z-square to an x-square, a y-square, and two identical parallelograms of sides x and y and smaller angles of $\theta - 90°$. Erwin Dintzl (1931) noted that two squares plus two parallelograms tessellate the plane in a manner that is a generalization of the tessellation of two squares in Figure 2.18. Dintzl's superposition of tessellations in Figure 4.25 explains Hunger's dissection (Figure 4.26), which is a generalization of Thābit's dissection in Figure 4.2.

Hunger's dissection requires eight pieces, so long as the parallelograms are not too skinny. Specifically, with $x \leq y$, an 8-piece dissection is possible if no angles in the parallelogram are smaller than $90° - \arcsin(x/(y\sqrt{2}))$. When the parallelogram does not satisfy this criterion, this superposition of tessellations yields a 10-piece dissection. But Dintzl superposed the same two tessellations so that each corner of the z-square is in the center of a y-square, as in Figure 4.27. The resulting dissection, in Figure 4.28, is then a generalization of Perigal's and uses nine pieces.

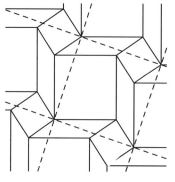

4.25: Dintzl tiling

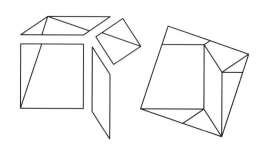

4.26: Hunger's law of cosines dissection

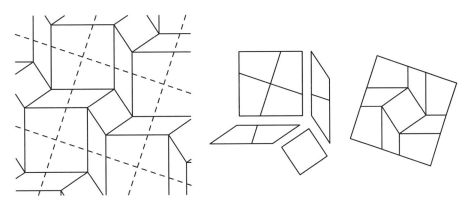

4.27: Perigal and Dintzl   **4.28:** Dintzl's law of cosines dissection

Erwin Dintzl was born in 1878 in Krems, Austria. He studied mathematics at the University of Vienna and was awarded a Dr. phil. in 1900.

In 1901 he started as a teacher in a Realschule in Vienna; in 1906 he became a professor at the kaiserliche königliche Erzherzog Rainer-Realgymnasium in Vienna. Dintzl's original area of research was number theory, but his interests shifted to the methodology of mathematics instruction. From 1925 to 1949 he also lectured at the University of Vienna on this topic. During the 1920s, he updated two textbooks of mathematical instruction for middle schools, and coauthored two volumes of mathematical exercises. Erwin Dintzl died in 1972.

CHAPTER 5

Triangles and Friends

With the bustle of Massachusetts Avenue behind him, Harry C. Bradley strode briskly along the gravel path laid parallel to the bank of the Charles River basin. He then pivoted 90° to face the Grand Court of the New Technology. The interconnecting stone buildings massed to form pavilions on either side as he approached the main portico of this new temple of progress. He was a professor at the premier technology school of a nation newly confirmed on the world stage. The dome in front of him capped the colonnaded center of these commandingly geometric buildings. In the few years since the move from the old "Boston Tech" on Boylston Street, he had responded to this realization of a dream with a constructive geometry of his own: transforming the hexagram into a square and imposing that cornerstone of Greek mathematics, the Pythagorean theorem, on that most fundamental of shapes, the equilateral triangle.

The year was 1922, and the place was the gleaming new campus of the Massachusetts Institute of Technology (MIT) in Cambridge, Massachusetts. The quiet professor of descriptive geometry felt justly proud of the result that he had sent off as a problem for the readers of the *American Mathematical Monthly* (*AMM*). He had turned around centuries of geometric practice by illustrating the Pythagorean theorem with equilateral triangles instead of squares. His dissection of triangles of side lengths x, y, and z graphically illustrated the relation $x^2 + y^2 = z^2$.

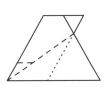

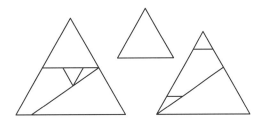

5.1: Hidden P-slide   **5.2:** Bradley's two triangles to one

For some unexplained reason, *AMM* readers had to wait eight years to see his 5-piece dissection. Bradley (1930) first formed a trapezoid by cutting a triangle off the top of the y-triangle and rotating it into position to yield a parallelogram with a triangle attached on the side. The dotted line in Figure 5.1 indicates an imaginary edge that separates the parallelogram on the left from the triangle on the right. He then applied a P-slide, shown with dashed edges, to the parallelogram. The dissection, shown in Figure 5.2, places the x-triangle on top of a trapezoid.

Following Bradley's lead, we first dissect equilateral triangles and then go on to hexagons and hexagrams. Just as there are two general methods for dissecting two squares to one square, there are two methods for dissecting two triangles to one. Harry Lindgren (1956) gave a second 5-piece method, shown in Figure 5.3. As in Figure 5.2, he converted the y-triangle to a trapezoid with base of length z. But he used the Q-slide technique to transform one quadrilateral to another with the same angles. Here the quadrilaterals are trapezoids, with the y-triangle viewed as a trapezoid with a top edge of length 0.

Lindgren (1956) introduced the *quadrilateral slide*, or *Q-slide*, and he discussed it in full generality in (1960). It transforms one quadrilateral to another with the

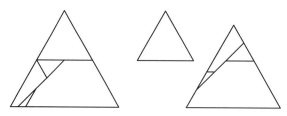

5.3: Lindgren's two triangles to one

same angles, as shown in Figure 5.4, where the quadrilateral ABCD on the left has a shorter base and thus is taller. Lindgren located point E on side AB so that line segment BE is the desired length of the edge on the right. He also located point F on side CD so that line segment FD is the desired length of the edge on the left. Finally, he let point G be the midpoint of line segment AE and H be the midpoint of CF.

He then made a cut from G to H, a second cut from E that is parallel to AD, and a third cut from F that is parallel to BC. He rotated the two triangles created by the second and third cuts 180° degrees, to match with a side of another piece. Then he slid the matched pieces along the line of the original cut, as in Figure 5.5. We have already seen how to hinge this dissection in Figure 3.13.

The dissection works for any set of dimensions within a certain range. Even with the most favorable shape for the quadrilaterals, the height of one is never more than three times the height of the other. When the height of one is precisely three times the height of the other, the quadrilaterals must be trapezoids, and in fact a 3-piece dissection exists in this case.

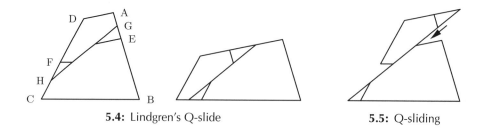

5.4: Lindgren's Q-slide   **5.5:** Q-sliding

The dissections in Figures 5.2 and 5.3 will work for any triangles, not just equilateral ones. For two similar isosceles triangles in which the corresponding sides are in the ratio $\tan(\theta)$, where θ is one of the two equal angles, Harry Hart (1877) found a 4-piece dissection.

Puzzle 5.1: Find a 4-piece dissection of a 1-triangle and a
$\sqrt{3}$-triangle to a 2-triangle.

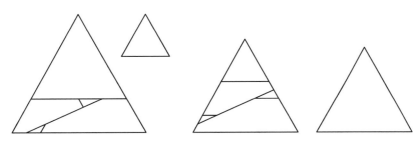

5.6: Two triangles to two triangles

Now let's dissect a pair of triangles to a different pair of triangles. Just as the corresponding dissection for squares (Figure 4.8) was a generalization of Perigal's dissection (Figure 4.6), my dissection in Figure 5.6 is a generalization of Lindgren's dissection in Figure 5.3. I cut a smaller triangle off the top of each of the larger triangles and convert the remaining trapezoid of one to the remaining trapezoid of the other via a Q-slide.

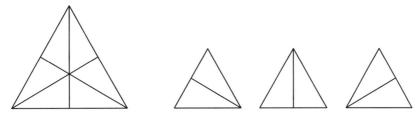

5.7: Three equal triangles to one

We can also dissect three triangles to one. Plato's *Timaeus* described the 6-piece dissection shown in Figure 5.7 for the case when all three triangles are equal. When the triangles are not equal, I identify two cases, each handled by an 8-piece dissection. Assume for the next two dissections that $x \le y \le z$. In the first case, $y \ge 2x$. Position the x-triangle flush against the y-triangle and perform a P-slide similar to that in Figure 5.1, giving a trapezoid. This takes four pieces, but works only when $y \ge 2x$, because of the restrictions on a P-slide. If we view the z-triangle as a trapezoid with a triangle on top, we can use a Q-slide to convert it to a longer and thinner trapezoid with the same triangle on top, in four pieces. We assemble

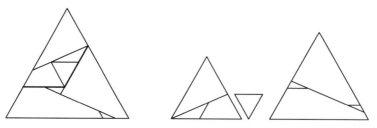

5.8: Three unequal triangles to one: I

the trapezoid from the x- and y-triangles and the figure from the z-triangle as in Figure 5.8 to give the w-triangle.

When $x \leq y \leq 2x$, we need a few more tricks. Position the x-triangle at the top of the w-triangle. We could convert the z-triangle to a trapezoid with a triangle on top and place it on the bottom of the w-triangle, but this would take four pieces. Instead of a trapezoid, I chose a quadrilateral whose boundary coincides with the slide line of the Q-slide, as suggested in Figure 5.11, so that I use only three pieces. This leaves a hole in the shape of a quadrilateral in the w-triangle. Then I convert the y-triangle to this quadrilateral in four pieces, using the TT2-strip technique, which we will see in Chapter 12, near the end of the discussion of the triangle-to-square dissection. The 8-piece dissection is in Figure 5.9.

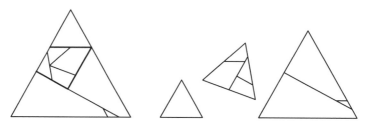

5.9: Three unequal triangles to one: II

For three triangles, as for three squares, there are special relationships that facilitate more economical dissections. These include relationships (4.2), (4.3), and (4.5).

For relationship (4.2), when $z = y$, I have found three cases. If $x \geq y$, then there is the 6-piece dissection in Figure 5.10. Place the two equal triangles flush against each other and slide them flush against the x-triangle. This gives an irregular-shaped piece that we can view as a triangle perched atop a trapezoid. Applying a Q-slide to the trapezoid compresses it to form a base for the triangle.

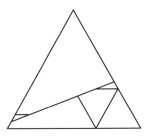
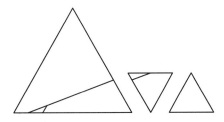

5.10: Three triangles to one, with two equal: I

If $y/2 < x < y$, then there is the 6-piece dissection in Figure 5.12. Position the y-triangle in the top corner of the w-triangle, and cut the z-triangle to form a quadrilateral, which we position flush against the y-triangle. If the quadrilateral were a parallelogram, then, as shown in Figure 5.11, the remaining portion would look like a trapezoid with a small triangle perched on top. Thus we can use the same trick that we used to convert the z-triangle in Figure 5.9. If we form a quadrilateral whose boundary coincides with the slide line of the Q-slide, we save a piece, giving the 6-piece dissection.

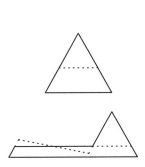

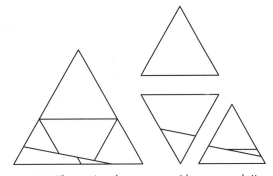

5.11: Trapping trapezoids   **5.12:** Three triangles to one, with two equal: II

If $x = y/2$, then there is a 4-piece dissection that makes a nice puzzle.

Puzzle 5.2: Find a 4-piece dissection of triangles for
$1^2 + 2^2 + 2^2 = 3^2$.

We can frame two puzzles from relationship (4.3), in which $x + y = w$. First, there is in general a 6-piece dissection. For the special case when $y = 3x/2$ and $z = \sqrt{3}x$, Dudeney (*Dispatch*, 1900b) gave a 5-piece dissection, with one piece turned over.

Puzzle 5.3: Find a 6-piece dissection of three triangles realizing $x^2 + y^2 + z^2 = w^2$, where $x + y = w$.

Puzzle 5.4: Find a 5-piece dissection of triangles for $2^2 + 3^2 + (\sqrt{12})^2 = 5^2$ in which neither the 2-triangle nor the 3-triangle is cut.

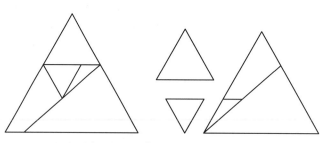

5.13: Three triangles: $w + x = y + z$

For relationship (4.5), in which $w + x = y + z$, I have found the 5-piece dissection shown in Figure 5.13. Position the x-triangle next to the z-triangle, with corners coincident. Then apply a P-slide, much as in Figure 5.1, to get a trapezoid with a top side of length y. Finally position the y-triangle on top.

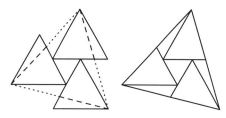

5.14: Wells's four triangles

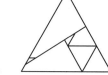

5.15: Four other triangles

An interesting case is that of four triangles to one, when three of the triangles are equal. By analogy to Abū'l-Wafā's dissection of squares in Figure 4.10, David Wells (1975) has found the 7-piece dissection in Figure 5.14. This dissection works whenever the one triangle that is unequal to any of the others is smaller. Robert Reid has suggested the puzzle in which the one triangle that is unequal to any of the others is larger. For this case, I have found a 7-piece dissection, shown in

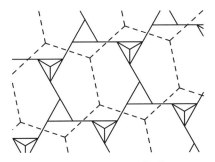

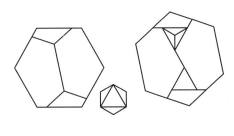

5.16: Hexagons overlaid: I

5.17: Lindgren's two hexagons to one

Figure 5.15. It uses a modified Q-slide and shares similarities with the dissections in Figures 5.10 and 5.12.

Hexagons are close friends of triangles because of their related structure. We start with hexagons for $x^2 + y^2 = z^2$. Let an x-hexagon denote a regular hexagon of side length x. Lindgren (1964b) found an 8-piece dissection for the case when $y > x\sqrt{3}$. As shown in Figure 5.16, he cut the x-hexagon into two equilateral triangles, which together with the y-hexagon form the tessellation element for the tessellation in Figure 2.16. He superposed this tessellation over the tessellation of hexagons from Figure 2.10. The resulting dissection is shown in Figure 5.17.

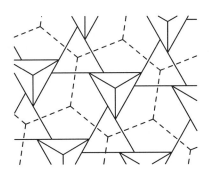

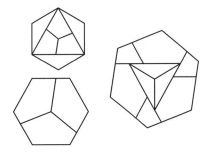

5.18: Hexagons overlaid: II

5.19: Lindgren's two hexagons to one

When $y < x\sqrt{3}$, Lindgren (1964b) gave a 9-piece dissection. In this case, the two triangles of the tessellation will not both fit inside the resulting hexagon. The tessellation of Figure 2.17 is used in place of the tessellation of Figure 2.16, and the superposition in Figure 5.18 is somewhat different. The resulting 9-piece dissection is shown in Figure 5.19.

For the case when $y = x\sqrt{3}$, Lindgren (1964b) gave a 6-piece dissection, which can be partially, variously hinged.

Puzzle 5.5: Find a partially, variously hinged 6-piece dissection of a 1-hexagon and a $\sqrt{3}$-hexagon to a 2-hexagon, such that the two smaller hexagons are each cut into three pieces.

I have found two interesting special cases for hexagons. Figure 5.20 shows a 6-piece dissection for $2^2 + 3^2 = (\sqrt{13})^2$. Figure 5.21 has a 6-piece dissection for $(\sqrt{7})^2 + 3^2 = 4^2$. The latter turns over two pieces.

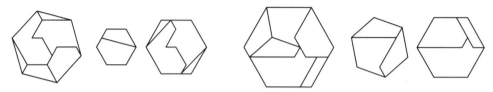

5.20: Hexagons: $2^2 + 3^2 = (\sqrt{13})^2$ **5.21:** Hexagons: $(\sqrt{7})^2 + 3^2 = 4^2$

Let's become ambitious and try hexagrams for $x^2 + y^2 = z^2$. Lindgren (1964b) described a clever method for dissecting two hexagrams to one. First, he converted the x- and y-hexagrams to rectangles in four pieces each: For each hexagram, he cut off the triangles corresponding to the top and bottom points of the star and nestled them into one of the side cavities. Then he made a vertical cut and swung the resulting piece around to complete the enclosure of the points. This approach is an extension of Bradley's hexagram element, which we will see in Figure 11.12.

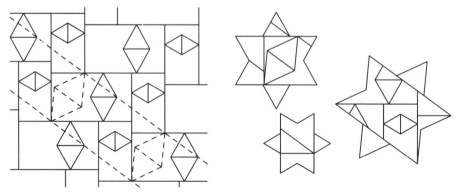

5.22: Hexagrams overlaid   **5.23:** Lindgren's two hexagrams to one

Next, Lindgren constructed a tessellation of the resulting rectangles for the x- and y-hexagrams, using the tessellation in Figure 2.19. Finally, he took a strip made from the element for the large hexagram and superposed it on the tessellation. There is flexibility in how to cut the hexagram element to give a rectangle. He made the cuts so that the pairs of triangles overlap with as little else as possible. Lindgren's tessellation and superposition are shown in Figure 5.22, and the 15-piece dissection is in Figure 5.23. This dissection is feasible as long as the pair of dashed triangles tilts enough to avoid crossing the vertical side of the rectangle for the y-hexagram. This holds whenever $x \leq y < x\sqrt{3}$.

As inspired as Lindgren's hexagram dissection is, it must also be classified in part as an opportunity lost. Lindgren did not determine the range of y/x for which his dissection is valid and thus did not determine what to do outside this range. For $y > x\sqrt{3}$, there is a dissection that is a show-stopper. My 13-piece dissection, shown in Figure 5.24, not only is translational but also has 6-fold rotational symmetry. Since the large star appears ready to give birth to the small one, I have named this dissection "Star Genesis."

Each small triangle in the y-hexagram has angles of 30°, 30°, and 120° and has a vertex coincident with a reflex vertex of the x-hexagram. We can apply the technique in Figure 5.24 to any star that is the union of a single type of rhombus. For example, the star {8/3} is the union of 45°-rhombuses. A 17-piece dissection is possible whenever $y > x\sqrt{2}$. Such stars are of the form {2n + 2/n}. What happens in the degenerate case, when $n = 1$? Treating a square as the union of four squares and noting that the triangles analogous to the 30°–30°–120° triangles in Figure 5.24 become 0°–0°–180° triangles and thus vanish, we get Perigal's dissection!

5.24: "Star Genesis": two hexagrams to one

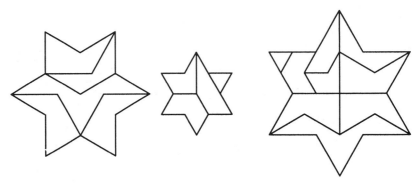

5.25: Reid's hexagrams for $1^2 + (\sqrt{3})^2 = 2^2$

In a manner similar to that of triangles and hexagons, there is an attractive dissection of an x-hexagram and a y-hexagram when $y = x\sqrt{3}$. I created a symmetrical 10-piece dissection, but it has been displaced by Robert Reid's 9-piece dissection. Shown in Figure 5.25, Reid's dissection turns a piece over. Will some clever reader find a 9-piece dissection with no pieces turned over?

My last technique doesn't cover hexagrams realizing $(\sqrt{3})^2 + 2^2 = (\sqrt{7})^2$, but Lindgren's does. However, I have found a more economical dissection than Lindgren's: the 13-piece solution shown in Figure 5.26. It pays a price for economy in that six pieces must be turned over, but it does enjoy 6-fold rotational symmetry.

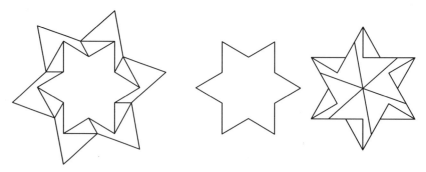

5.26: Hexagrams for $(\sqrt{3})^2 + 2^2 = (\sqrt{7})^2$

CHAPTER 6

All Polygons Created Equal

Seated humbly on the ground, with writing paper bal-
anced on my knees, I faithfully transcribe the words of
my master. Praise to Allah, I am blessed to be a disci-
ple of the renowned Abū'l-Wafā at the caliph's palace
school. The great House of Wisdom here in Baghdad
stores amidst its treasures his scholarly commentaries on
mathematics. From his lips I am honored to learn the ge-
ometry of the Greeks. And he holds us spellbound with
his own remarkable constructions: How wondrous are
his divisions of squares that in their unity form a large
square! The master reminds us that this is the fruit of
many centuries and many civilizations. It is so perfect; I
do not think it can ever be surpassed.... Now how am I
to comprehend that last construction?

Alas, they don't write instructor evaluations like they used to, at least if you would
believe this counterfeit passage, masquerading as the musings of an Arabian student in tenth-century Baghdad. More authentic than these supposed musings of an
adolescent mind is the role that Arabic-Islamic civilization played in the preserva-
tion and extension of the learning of the Greeks, at a time when European civilization was in eclipse. The dissection of squares referred to was but a tiny part of the
remarkable activity carried on not only in mathematics but also in the many other
disciplines that flowered during that period. The dissection is beautiful, and the
natural one with which to start a chapter on assembling identical copies of regular
polygons into a larger copy of themselves.

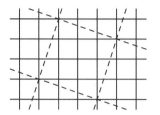
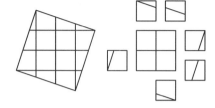

6.1: Tessellating squares   **6.2:** Abū'l-Wafā: ten squares to one

Abū'l-Wafā's method for dissecting squares works whenever the number of identical squares is $a^2 + b^2$, where a and b are integers. We generate his dissection by superposing two tessellations of squares, where the large squares have area $a^2 + b^2$ times that of the smaller squares. The side of a large square corresponds to the hypotenuse of a right triangle with legs of length a and b. For the case of $a = 1$ and $b = 3$, which corresponds to ten squares to one, the tessellations are in Figure 6.1, and the dissection is in Figure 6.2. The number of pieces is $a^2 + b^2 + 2(a + b - \gcd(a, b))$, where $\gcd(a, b)$ is the greatest common divisor of a and b.

As pretty and as natural as the preceding approach is, it can sometimes be surpassed. In particular, the example of thirteen squares to one is not minimal. Robert Reid has found the 20-piece dissection in Figure 6.3. Reid filled out a large portion of the large square with uncut squares, aligned with its sides. Then he filled the margins with rectangles cut from other squares and used a P-slide to finish off the final corner.

Abū'l-Wafā's method also leads to economical dissections of $a^2 + b^2$ squares of one size to $c^2 + d^2$ squares of another size. First superpose an initial tessellation of squares and a second tessellation of squares, each of whose area is $a^2 + b^2$ times

Abū'l-Wafā al-Būzjanī was a Persian who was born in 940 in the province of Khurasan (in present-day Turkmenistan) and who died in Baghdad in 998. He was a renowned mathematician and astronomer, who as associated with the Bait al-Hikmah (House of Wisdom) in Baghdad. As a mathematician, he was one of the last Arabic translators and commentators of the Greek writers. He wrote commentaries on the algebra of Hipparchus and the algebra of Muhammed ibn Musa al-Khwārizmī of which only the titles survive; translatd and extended Diophantus; made various contributions in trigonometry; wrote a textbook on practical arithmetic (for scribes and businessmen); and wrote a book of geometrical constructions. Fragments of this latter book survive in the form of a Persian translation of an Arabic text based on an abridgment of Abū'l-Wafā's lectures that was prepared by a talented disciple.

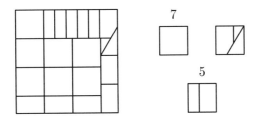

6.3: Reid's dissection of thirteen squares to one

that of the first. Then superpose a third tessellation of squares, each of whose area is $c^2 + d^2$ times that of the first. Actually, we do not need the first tessellation and leave it out of Figure 6.4, which gives the tessellations for five squares to two. The 12-piece dissection is given in Figure 6.5. David Paterson (1995a) surveyed dissections for n squares to m, for many combinations of n and m, as has Robert Reid.

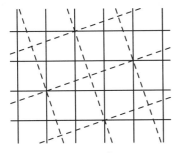

6.4: Squares and squares

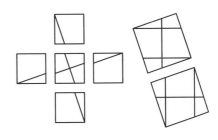

6.5: Five squares to two

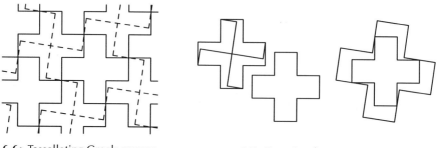

6.6: Tessellating Greek crosses   **6.7:** Two Greek crosses to one

The tessellation approach works well even when the figure to be dissected is not a square, but forms a tessellation element with the repetition pattern of a square. Since a Greek Cross satisfies the preceding criterion, we are blessed with economical dissections of $a^2 + b^2$ Greek Crosses to one. Dudeney (*Dispatch*, 1900c) gave a 5-piece dissection of two crosses to one. The superposed tessellations are in Figure 6.6, and the dissection itself is in Figure 6.7. This dissection can be partially hinged. Lindgren (1964b) gave a dissection of five Greek Crosses to one.

Puzzle 6.1: Find a dissection of five Greek Crosses to one.

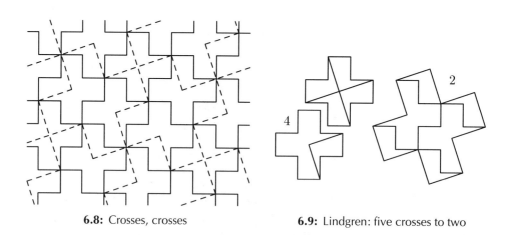

6.8: Crosses, crosses   **6.9:** Lindgren: five crosses to two

As with squares, there is an economical dissection of five Greek Crosses to two. The superposition of tessellations in Figure 6.8 leads to the 12-piece dissection by Lindgren (1964b) in Figure 6.9.

Harvesting the fruit of Abū'l-Wafā's civilizations, Lindgren (1964b) identified a similar approach for dissecting triangles. His method, based on tessellations of triangles, gives economical dissections of $a^2 + ab + b^2$ triangles to 1. The side of the large triangle corresponds to the side opposite a 120° angle in a triangle whose other sides have length a and b. Paterson (1989) showed that the number of pieces is $a^2 + ab + b^2 + 3(a + b - \gcd(a, b))$. We have already seen a 6-piece dissection of three triangles to one in Figure 5.7. The superposition of the tessellations is shown on the left in Figure 6.10. With $a = 1$ and $b = 2$, there is a 13-piece dissection of seven triangles to one, given by Hugo Steinhaus (1960). The superposition of the tessellations is shown on the right in Figure 6.10.

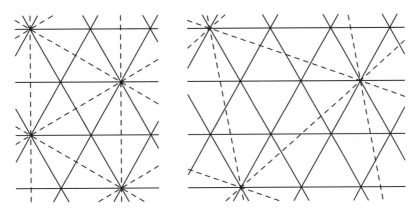

6.10: Tessellations for triangles: three to one, seven to one

Although Lindgren's approach for triangles is as nice as Abū'l-Wafā's for squares, it suffers from the same deficiency: It can sometimes be surpassed. In unpublished work, David Collison provided the spoiler: a 12-piece dissection of seven triangles to one. We see a modified version of Collison's dissection in Figure 6.11. He converts three of the triangles to a trapezoidal base, upon which rests a triangle formed from the remaining four triangles. To form the base, he cuts two triangles horizontally and cuts a third triangle to fill in the remaining portion, which is a trapezoid with a triangle on top, similar to that in Figure 5.11. A Q-slide transforms one to the other, as in Figure 5.12.

David Collison's dissection suggests an approach that should work well for any sufficiently large number of triangles dissected to one large triangle: Pack as many triangles as possible into one corner, leaving a long, narrow trapezoidal region, and then cut the remaining triangles to fill this region. I have not tested this method for large numbers of triangles.

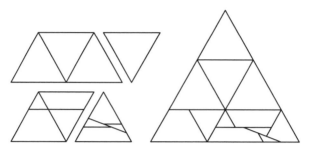

6.11: Collison's seven triangles to one

But this method, too, can be surpassed. For the case of five triangles to one, Alfred Varsady (1989) produced an ingenious 9-piece dissection (Figure 6.13). It results from the 6-piece dissection of triangles for $x^2 + y^2 = z^2$ (Figure 6.12). Position the y-triangle flush above the x-triangle, with their left vertices coincident. Draw the z-triangle (in dashed lines), with one vertex coincident with the lowest vertex of the x-triangle, and its top edge bisecting the right side of the y-triangle. The cuts induced by this superposition yield pieces that fill the vacant spaces, with two pieces turned over. It is humbling to wonder how this dissection was discovered. Returning to his dissection of five triangles to one, Varsady takes one triangle to be the x-triangle and forms the y-triangle from the other four.

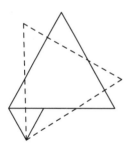

6.12: Two triangles to one

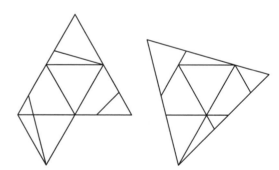

6.13: Varsady's five triangles to one

Lindgren (1964b) identified tessellation dissections for $a^2 + ab + b^2$ triangles to $c^2 + cd + d^2$ triangles. Paterson (1989) displayed a variety of these, including a 21-piece dissection of seven triangles to three and a 33-piece dissection of thirteen triangles to three. But even here the tessellation-based approach can be surpassed. In a manner suggested by Figure 6.12, I found the 27-piece dissection of thirteen triangles to three shown in Figure 6.14. For each of the three large triangles, take four of the smaller triangles together as the y-triangle, and use a third of the

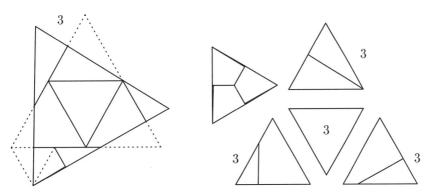

6.14: Thirteen triangles to three

thirteenth small triangle for the *x*-triangle. Cut a thin sliver off of the third and swing it around. The slivers are so thin that they appear in the figure only as a thickening of the line segments. The dissection has 3-fold replication symmetry, but several pieces must be turned over.

Harvesting more of Abū'l-Wafā's fruit, Lindgren (1964b) also identified a tessellation-based method that gives economical dissections of $a^2 + ab + b^2$ hexagons to one. Superposed tessellations for a 6-piece dissection of three hexagons to one are shown on the left in Figure 6.15. With $a = 1$ and $b = 2$, there is a 12-piece dissection of seven hexagons to one, whose superposed tessellations are shown on the right in Figure 6.15.

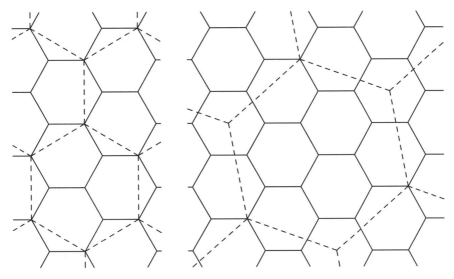

6.15: Tessellations for hexagons: three to one, seven to one

With its relative abundance of sides, the hexagon may be one polygon for which no other approach does better than the tessellation method, when it applies. But it doesn't work for five-to-one dissections of hexagons. Alfred Varsady recognized that an approach related to his approach for triangles would work nicely. I refined his unpublished 18-piece dissection, giving the 16-piece dissection shown in Figure 6.16. We can derive this by making six copies of Figure 6.12 and arranging them so that the small triangles all share a central point. The resulting six triangles form the outline of the hexagon. Once we have established the outline, a more careful choice of cuts gives the result. Six pieces are turned over in this dissection.

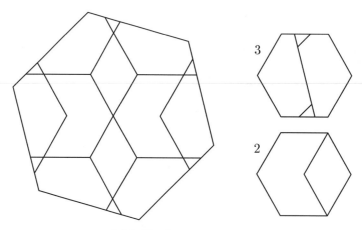

6.16: Five hexagons to one

Inspired by Abū'l-Wafā, we have investigated figures that tessellate the plane, though some of the best dissections do not rely on tessellations. Since pentagons cannot tile the plane, are they immune to the techniques in this chapter? No, they aren't, for reasons that we shall explore in Chapter 10. Varsady (1989) identified a neat 21-piece dissection of six pentagons to one (Figure 6.17). He centered one small pentagon in the large pentagon and cut each remaining pentagon along a diagonal into an isosceles triangle and a trapezoid. He then cut each of these pieces once again. The dissection has 5-fold rotational symmetry.

To comprehend this construction, we perform it in two steps. In Figure 6.18, we arrange the five isosceles triangles around the center pentagon, using dashed line segments to connect the apexes. The dashed lines indicate cuts that are similar in function to those used in Abū'l-Wafā's dissection of three squares in Figure 4.10. We then surround the resulting pentagon by the five trapezoids, as in Figure 6.19.

Dashed line segments once again indicate cuts that have a function similar to the cuts in Abū'l-Wafā's dissection. In Chapter 10 we will discover that this dissection uses tessellations, but not tessellations of the plane!

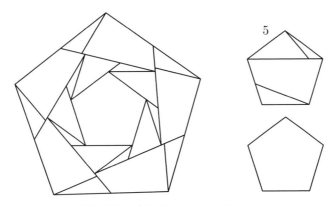

6.17: Varsady's six pentagons to one

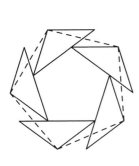

6.18: Adding isosceles

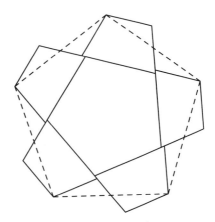

6.19: And trapezoids too

CHAPTER 7

First Steps

Down two levels in the sub-basement, the librarian un-
locks first the door to the special collection and then
the cabinet door. Retrieved from within is a substantial
leather-bound volume, the third in the collected works
of Girolamo Cardano, containing De Rerum Varietate.
The thick pages are in good condition, save for the bleed-
through of the print. The volume has survived for over
three centuries, outlasting so many, many people. And
before its printing in 1663, its contents had survived
a century from their penning in 1557. Cached amidst
a variety of things is an illustration of how to cut a
(2a x 3b)-rectangle into just two pieces that rearrange
into a (3a x 2b)-rectangle. The Latin text now obscures
an explanation. Written by a man who embraced the
then-new notion of shared knowledge, his contribution
has endured to lie buried in a basement, locked within a
room, within a cabinet, within a now-dead language.

The dissection Cardano (1663) described is an example of the *step* technique, so named because of the resemblance to stair steps. In its simplest form, the step technique cuts a rectangle in a zigzag pattern, alternating horizontal cuts of one length with vertical cuts of another. By shifting the two resulting pieces by one step relative to each other, we form a different rectangle. Figure 7.1 illustrates the case for converting a $(3a \times 4b)$-rectangle to a $(4a \times 3b)$-rectangle. Besides giving this case, Cardano also converted a $(2a \times 3b)$-rectangle to a $(3a \times 2b)$-rectangle. For those who never learned their Latin, a discussion of the technique is also in

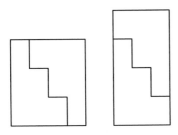 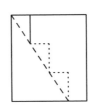 

7.1: A step dissection   **7.2:** Conversion from P-slide

(Dudeney 1926a). Dissections that use the step technique are examples of rational dissections.

There is a neat relationship between the step technique and the P-slide technique. When the length of the base of the triangle in the P-slide evenly divides the length of the corresponding side of either rectangle, then we can convert the P-slide to a step. This conversion is illustrated in Figure 7.2, with dashed lines indicating cuts in the P-slide that we delete, and dotted lines indicating cuts that we introduce. We save a piece in this case. We can view a number of dissections in the next chapter as resulting from this conversion applied to a corresponding dissection in Chapter 4.

Girolamo Cardano (Jerome Cardan) was an Italian physician and mathematician whose remarkable and turbulent life extended from 1501 to 1576. He received the doctor of medicine from the University of Padua but was denied admission into the College of Physicians in his hometown of Milan on the pretext that he was illegitimate. Only after publishing a book containing both a large dose of medical good sense and a virulent attack on the profession was he finally admitted. As his fame grew, he was offered, and turned down, several positions as court physician. But he could not refuse 1800 gold crowns to cure the archbishop of Scotland of asthma.

Cardano became well known in all of Europe because of his popular books on science and philosophy. His collected works filled ten oversize volumes. Mathematics was a second vocation, and his book on algebra profoundly influenced the development of European mathematics. Having cajoled out of the mathematician Tartaglia the solution method for cubic equations of the form $x^3 + ax + b = 0$, Cardano and his associate Ferrari discovered the solution methods for general cubic and quartic equations. Going further, he determined the number of roots an equation may have and recognized the need for not only negative but also complex numbers to obtain complete solutions.

In this chapter we investigate rational dissections of a pair of squares to one square. We have already seen two different general approaches in Figures 4.2 and 4.6 for dissecting two squares to one square in five pieces. So we search here for dissections that need only four pieces. Four pieces is the best we can do: Since the side length of the resulting square is greater than the side lengths of either of the given squares, no single piece can contain two corners of the resulting square. Also, we shall look for 4-piece dissections of squares that result in 3-piece dissections when the two squares are attached.

Let x, y, and z be the side lengths of the squares. Diophantus, a fourth-century Greek mathematician from Alexandria, found a method to generate all basic solutions to $x^2 + y^2 = z^2$. His method amounted to the following, which the seventh-century Indian mathematician Brahmagupta stated explicitly: Every basic solution is of the form $x = m^2 - n^2$, $y = 2mn$, and $z = m^2 + n^2$, where m and n are relatively prime and $m + n$ is odd.

There are two classes of solutions to $x^2 + y^2 = z^2$ that seem to be special. Pythagoras identified the first class, for which $m = n + 1$ in Diophantus's method. This class contains such familiar solutions as $3^2 + 4^2 = 5^2$, $5^2 + 12^2 = 13^2$, and $7^2 + 24^2 = 25^2$. Henry Dudeney and Sam Loyd gave 4-piece rational dissections for these three solutions. These are shown in Figures 7.3–7.6.

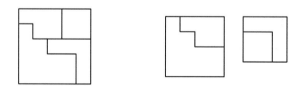

7.3: Loyd's squares for $3^2 + 4^2 = 5^2$

The step technique appears in Figures 7.3, 7.5, and 7.6. In particular, the piece in the upper left corner of the 5-square of Figure 7.3, the pieces in the lower left and upper right corners of the 13-square of Figure 7.5, and the piece in the upper right corner of the 25-square of Figure 7.6, all conform to the step technique.

The dissection in Figure 7.3 is just one of many that illustrate $3^2 + 4^2 = 5^2$. It was given as the solution to Sam Loyd's "Guido Mosaics Puzzle," which originally appeared in (Loyd, *Inquirer*, 1901b). The puzzle demonstrates Loyd's flair for dressing up puzzles. Notice how seamlessly Loyd incorporates into the puzzle statement the rationale for requiring a rational dissection and excluding any rotation:

THE GUIDO MOSAICS PUZZLE BY SAM LOYD

PROPOSITION—Show how to divide the mosaics into two squares.

It is not generally known that the celebrated piece of Venetian mosaic by Dome-chio, known as the Guido collection of Roman heads, was divided originally into two square groups, which were discovered at different periods. They were brought together and restored into what is supposed to be their correct form, in 1671.... We will reverse the problem, and ask you to divide the large square into the fewest number of pieces which can be refitted into two squares which conform to the conditions described.... It is assumed that we must cut on the lines only, so as not to destroy the heads.... As a lesson in puzzle construction it may be said that problems of this kind which call for the "best" answers, in the "fewest number of pieces," etc., offer great scope for cleverness. Anyone might find a solution in many pieces, or which stands some of the old Romans upside down.

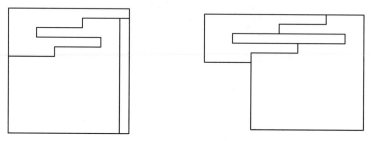

7.4: Stutter-step applied to $5^2 + 12^2 = 13^2$

Loyd's dissection for $5^2 + 12^2 = 13^2$ is shown in Figure 7.5. It appeared originally in (Loyd, *Inquirer*, 1899a) with the 5-square and 12-square attached, thus requiring just three pieces. Again, there are numerous possible dissections, so that Loyd dressed it up as "Mrs. Pythagoras' Puzzle" to exclude several possibilities. These included the dissection in Figure 4.4 and a dissection by Dudeney (*London*, 1902b) that rotates a piece (Figure 7.4).

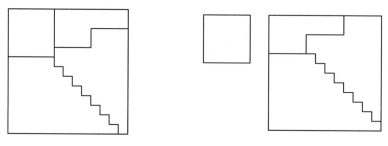

7.5: Loyd's squares for $5^2 + 12^2 = 13^2$

Mrs. Pythagoras took counsel with her spouse regarding the best way of squaring a two-square formed remnant of Athenian matting.... "Now, Thag," says she, for she always called him that in the house, "I am feared these goods will fray if they are cut on the bias, so I want to get along without that hippopotamus line. Here is a plan which will also do it in three pieces: [Figure 7.4 but with the two smaller squares attached].... "But, I don't like it altogether, Thag; you see the pattern don't run quite right on the squares in that long piece. Can't you find a perfect answer without giving any of the squares that half turn? I know it can be done."

The puzzle of dissecting squares corresponding to $7^2 + 24^2 = 25^2$ was an instance in which Dudeney was bettered. He had posed the puzzle in his "Perplexities" column and had supplied a 5-piece dissection in (Dudeney, *Strand*, 1924a). But he then received a 4-piece dissection from a Sgt. E. T. Richards, which he published in (Dudeney, *Strand*, 1926a).

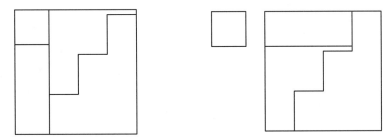

7.6: Sgt. E. T. Richards's squares for $7^2 + 24^2 = 25^2$

A technique that is related to, but less common than, the step technique is what we shall call a *stutter-step*. The stutter-step is a compound cut that has first a steplike zigzag, then a cut in the reverse direction, and finally another steplike zigzag. When we shift the two resulting pieces by one step relative to each other, we open up a rectangular hole, which we then fill by a rectangle of the appropriate dimensions. Dudeney (1907) used this technique in the dissection in Figure 7.4. In this dissection a piece consisting of a 5-square attached to a 12-square is cut into three pieces in a way that preserves a checkerboard pattern.

An earlier example of the stutter-step is used in the solution to the A&P Baking Powder Puzzle. This puzzle appeared on an 1885 advertising card that has been attributed to Sam Loyd by Jerry Slocum (1996). Given a (23×15)-rectangle with a (7×3)-rectangular hole in the middle, the puzzle is to cut the figure into two pieces that form a square. An illustration of the puzzle and its solution is given in Figure 7.7. The original puzzle card supplied only the outline of the rectangle and

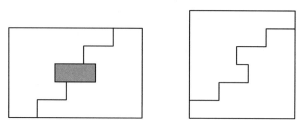

7.7: Loyd's A&P Baking Powder Puzzle

the hole; a significant portion of the puzzle must have consisted of identifying exact dimensions from which to work. Dudeney (*Strand*, 1926c) also used a stutter-step in the solution to this next puzzle.

Puzzle 7.1: Given a (40×33)-rectangle with an (8×3)-rectangular hole cut out of it, cut it into two pieces that give a 36-square.

Having taken our first few steps, are we now ready for a hike up the mountain? Can we find a general method for dissecting any set of squares in Pythagoras's class? Although little is suggested by the dissection in Figure 7.3, the dissections in Figures 7.5 and 7.6 are easy to generalize. David Collison (1980) appears to have been the first to obtain a general approach for Pythagoras's class. Although Collison did not give an example of his method for squares, the dissection in Figure 7.8 seems to conform to his approach, which we discuss in Chapter 9. I describe here a different approach, Method 1A, based on Figure 7.5. In its description, n is the value in Diophantus's formula. I will extend this approach to handle two related puzzles in the next chapter.

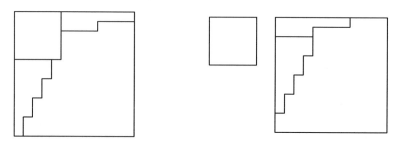

7.8: A Collison-like approach for $5^2 + 12^2 = 13^2$

Any dissection produced by Method 1A is translational. If the x- and y-squares are attached, then a trivial modification of Method 1A gives a 3-piece dissection. Loyd's dissection in Figure 7.5 illustrates Method 1A for the value $n = 2$, which corresponds to $5^2 + 12^2 = 13^2$. Since the dissections need no rotation, we can use the method for similar rhombuses whose side lengths satisfy the required relationship.

Method 1A: For squares in the class of Pythagoras.

1. Cut an x-square from the upper left corner of the z-square.

2. Step starting 1 left of the lower right corner of the z-square:
  Move/cut up a distance of 1.
  $[2n^2 - 1$ times]: {Move/cut left 1; Move/cut up 1.}

3. Step starting 2 below the upper right corner of the z-square:
  Move/cut left a distance of $2n$.
  $[n - 1$ times]: {Move/cut down 2; Move/cut left $2n$.}

Puzzle 7.2: Give a general rule for all members of Pythagoras's class, based on the dissection in Figure 7.6.

Returning from our hike up "Mt. Pythagoras," we ask whether there is another class of solutions to $x^2 + y^2 = z^2$ for which 4-piece dissections are possible. A likely candidate for this second class is the one identified by Plato, which has $n = 1$ in Diophantus's method. The first three solutions in Plato's class are $3^2 + 4^2 = 5^2$, $15^2 + 8^2 = 17^2$, and $35^2 + 12^2 = 37^2$. Harry Lindgren (1964b) gave a 5-piece dissection for $15^2 + 8^2 = 17^2$ and claimed that 4-piece dissections were possible only for Pythagoras's class. However, I have found that Lindgren's claim is in error. My Method 2A gives a 4-piece dissection not only for $15^2 + 8^2 = 17^2$ but for any member of Plato's class.

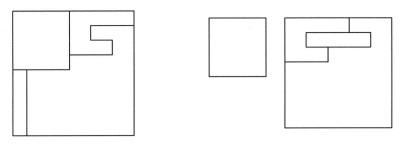

7.9: Method 2A applied to $15^2 + 8^2 = 17^2$

Method 2A: For squares in the class of Plato.

1. Cut a y-square from the upper left corner of the z-square.

2. Cut a $(2 \times (m-1)^2)$-rectangle from the lower left of the z-square.

3. Stutter-step starting 2 below the upper right corner of the z-square:
> $[m/2 - 1$ times$]$: {Move/cut left $2(m - 1)$; Move/cut down 2.}
> Move/cut to the right a distance of $(m - 3)(m - 1)$.
> $[m/2 - 1$ times$]$: {Move/cut down 2; Move/cut left $2(m-1)$.}

We illustrate Method 2A in Figures 7.9 and 7.10 for the values $m = 4$ and $m = 6$, which correspond to $15^2 + 8^2 = 17^2$ and $35^2 + 12^2 = 37^2$. Although I rotated the rectangular piece 90° with respect to the other pieces, I did center it horizontally in the y-square. If we define a distance of -1 to the right as equivalent to a distance of 1 to the left, then when $m = 2$, Method 2A gives a dissection of $3^2 + 4^2 = 5^2$ in which we do not cut the 4-square.

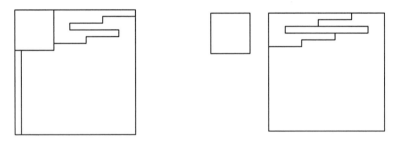

7.10: Method 2A applied to $35^2 + 12^2 = 37^2$

Because the step technique is centuries old, it is newsworthy that the world of business and technology has just stumbled onto it. In March 1995, IBM introduced its ThinkPad 701C minilaptop computer. When you lift the lid to expose the screen and the keyboard, there is a surprise. The keyboard separates into two pieces, the top piece sliding down and to the right, and the bottom piece sliding up and to the left. With the lid fully open, the two halves of the keyboard snap back together into a shape that is longer and narrower. Sticking out an inch beyond the case on each side, the keyboard is suggestive of wings, prompting the nickname "butterfly." You find yourself opening and closing the lid repeatedly just to watch this movement!

The keyboard is shown in the open-lid configuration in Figure 7.11, and in the closed-lid configuration in Figure 7.12. The solid lines indicate the boundaries of the two pieces. The keys on a standard keyboard occupy an $11\frac{1}{4} \times 6$ inch rectangle. If we could just nudge some of the keys a little to the right or the left, then we could

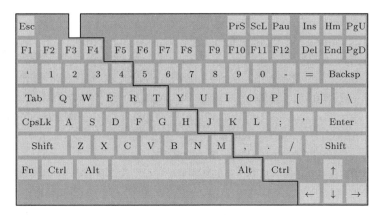

7.11: Karidis's keyboard – open …

convert the $9 \times 7\frac{1}{2}$ inch rectangle to the desired shape by a simple step dissection. But preserving the actual key position makes life much harder, since the second row of letters is offset from the first row of letters by $\frac{3}{8}$ inch, which is one-half the width of a letter's key; the third row is offset to the left from the second row by $\frac{3}{16}$ inch; the fourth row is offset from the third row by $\frac{3}{8}$ inch once again. The solution is to alternate steps of length $1\frac{1}{8}$ inches with those of length $\frac{15}{16}$ inch. When the pieces are shifted, the piece on the upper right moves up two steps with respect to the piece on the lower left. The resulting dissection is not of one rectangle to another, although this fact is not readily apparent when handling the laptop. The "notches"

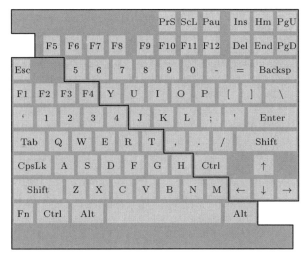

7.12: And shut!

in Figure 7.12 are hidden by a closed lid, and the notch in the upper left corner of Figure 7.11 is hidden by a plastic flap. For a real-world problem, you have to design a real-world solution!

IBM seems to have been the first company to apply dissection techniques in the design of a major product. Surprisingly, the designer at IBM knew nothing of past work in dissections. And it took less than two years from the moment when the inspiration struck the mechanical engineer John Karidis until the computer (with many other advanced features) was introduced.

Karidis had been pondering the problem of how to fit a full-sized keyboard into a case whose length and width are the same as the screen's. His experiments with origami hadn't been particularly helpful. Then he realized that the keyboard and screen areas were roughly equal, even though the keyboard was longer and thinner. Soon he thought of two of his three-year-old daughter's triangular wooden blocks sliding past each other. He raced to the copy machine, made copies of a keyboard, and spent the afternoon cutting apart the copies and rearranging the pieces. His demonstration at a meeting soon afterward, using a Plexiglas block mockup, caught the attention of IBM managers, and the pieces soon fell into place.

CHAPTER 8

Step Right Up!

In Puzzles and Curious Problems, *a posthumous collection of puzzles by the English puzzlist Henry E. Dudeney, the following problem appears: "A cabinetmaker had a perfect square of beautiful veneer which he wished to cut into six pieces to form three separate squares, all different sizes. How might this have been done without any waste?"*

A pretty puzzle, and yet in its framing the master's mortising chisel had slipped: Just five pieces suffice to form a set of three separate squares. And the number of such sets would fill a large warehouse. Even Dudeney's prototype set requires no more than five pieces, as Dudeney's old American rival Sam Loyd had catalogued more than two decades earlier! Furthermore, displaying his own consummate craftsmanship, Loyd had even preserved a checkerboard pattern in his dissection.

Dudeney (1931b) based his 6-piece dissection on the smallest numerical identity that satisfies the specified requirements, namely, $2^2 + 3^2 + 6^2 = 7^2$. Loyd (*Home*, 1908a) gave a 5-piece dissection for the same identity, shown with its checkerboard pattern in Figure 8.1.

This puzzle is remarkably rich even when we ask for 5-piece dissections. We needn't focus just on the case of $2^2 + 3^2 + 6^2 = 7^2$, since there are an infinite number of integral solutions to $x^2 + y^2 + z^2 = w^2$ that have 5-piece dissections. We shall find that the same situation exists for $x^2 + y^2 = z^2 + w^2$.

When you step into our warehouse full of cabinets, be ready to be ambushed by our friendly furniture salesperson: "A hearty welcome to our showroom warehouse,

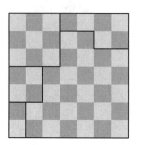
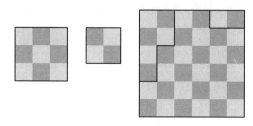

8.1: Loyd's squares for $2^2 + 3^2 + 6^2 = 7^2$

folks! You have come to the home of the square deal, with all the latest in $x^2 + y^2 + z^2 = w^2$ cabinetry in stock. Just remember our motto: 'Square Cabinets for Square Living.' But before we walk you through our showroom warehouse, have you heard of our special line of $x^2 + y^2 = z^2 + w^2$ cabinets?"

Let $z > \max\{x, y\} \geq \min\{x, y\} > w$. We saw a 6-piece dissection that works for all values (not just integers) in Figure 4.8, so we identify special cases for which five pieces suffice. No rational dissection can do better: Since z is strictly greater than x or y, each corner of the z-square must be in a different piece, and, in addition, the w-square contributes at least one piece. If the squares are attached to each other, then a general 4-piece dissection is possible, as suggested by Figure 4.9, so in that case the goal is to find a 3-piece dissection.

Sam Loyd's letterhead in the early 1900s.

A method for finding all integral solutions to $x^2 + y^2 = z^2 + w^2$ is implied by a specific derivation of $7^2 + 4^2 = 8^2 + 1^2$ by Diophantus. The formula was stated explicitly and proved by the Italian mathematician Leonardo of Pisa (Fibonacci) in 1225. Choose positive integers m, n, p, and q, such that $n < m$, $q > p$, and $mp \neq nq$. Then *Fibonacci's formula* is

$$x = mp + nq, \qquad y = mq - np, \qquad z = mq + np, \qquad w = |mp - nq|$$

Both $x^2 + y^2$ and $z^2 + w^2$ equal $(m^2 + n^2)(p^2 + q^2)$. When $p = n$ and $q = m$, Fibonacci's formula reduces to Diophantus's formula for generating solutions to $x^2 + y^2 = z^2$.

A nice surprise is that there are two classes of solutions that are closely related to the classes for $x^2 + y^2 = z^2$ in the preceding chapter, and the dissection methods are also closely related. I connect these classes by the following identity. For any integer a, we have

$$a^2 + (a - 3)^2 = (a - 1)^2 + (a - 2)^2 + 2^2 \tag{8.1}$$

Samuel Loyd, better known as **Sam Loyd**, was born in Philadelphia, Pennsylvania, in 1841. He attended public schools in New York until age seventeen and studied to be a civil engineer. Concentrating on chess from a young age, Loyd became world-renowned as a composer of chess problems. In 1860 he was engaged by the *Chess Monthly* as a problem editor. In 1878 he published *Chess Strategy*, a collection of many of his best chess problems.

In the early 1870s Loyd began to focus more on mechanical puzzles used as advertising giveaways, of which he was an unmatched inventor. The first of these was "The Trick Donkeys," which P. T. Barnum distributed by the millions. Loyd popularized the 14-15 puzzle, which became a national craze. (He claimed to have invented the puzzle and to have offered a $1,000 prize for a correct solution. There was no solution, a fact that he claimed prevented him from getting a patent, since there was no "working model" for it.) His "Horse of Another Color" puzzle sold millions. Loyd also claimed to have invented the popular game "Parcheesi." In 1896 Loyd patented his famous "Get Off the Earth" puzzle, in which one of thirteen Chinese warriors is made to vanish. Two later variations were "The Lost Jap" and "Teddy and the Lions."

Starting in the 1890s, Loyd published mathematical puzzle columns in a variety of newspapers and magazines. A number of his best puzzles were collected into four issues of *Our Puzzle Magazine*, which were combined after his death in 1911 into Loyd's *Cyclopedia of Puzzles*.

When we take an appropriate value for a, and then substitute in $x^2 + y^2 = z^2$, we generate an equation for a new class.

For the first class, start with $x^2 + (z-1)^2 = z^2$, which gives solutions in the class of Pythagoras. Take $z = a$ and $x = b$, and then substitute for a^2 in equation (8.1), giving $(a-3)^2 + b^2 = (a-2)^2 + 2^2$. Here b and a take the values of x and z, respectively, in the class of Pythagoras, so that I call this class the *Pythagoras-minus class*. It includes $10^2 + 5^2 = 11^2 + 2^2$, and $22^2 + 7^2 = 23^2 + 2^2$. We can express the class in terms of Fibonacci's formula, by taking $m = n + 1$, $p = n + 2$, and $q = n + 3$. Method 1B handles this class in a fashion similar to Method 1A from Chapter 7.

Method 1B: For squares in the Pythagoras-minus class.

1. Cut a y-square from the upper left corner of the z-square.

2. Step starting 1 left of the lower right corner of the z-square:
   Move/cut up a distance of 1.
   $[2n^2 + 4n - 1$ times]: {Move/cut left 1; Move/cut up 1.}

3. Step starting 2 below the upper right corner of the z-square:
   Move/cut left a distance of $2n$.
   $[n$ times]: {Move/cut down 2; Move/cut left $2n+2$.}

A dissection produced by Method 1B is translational. If the two smaller squares are attached, then we have a 3-piece dissection. Figure 8.2 gives the dissection for $n = 1$, which corresponds to $10^2 + 5^2 = 11^2 + 2^2$.

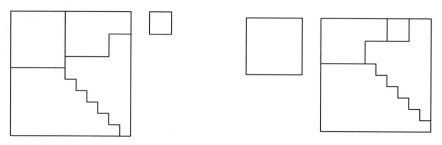

8.2: Method 1B applied to $10^2 + 5^2 = 11^2 + 2^2$

Let's derive a second class from $(z-2)^2 + y^2 = z^2$, which includes all solutions in Plato's class, as well as twice each solution in Pythagoras's class. Take $z = a$ and

$y = b$ and then substitute for a^2 in equation (8.1), giving

$$b^2 + (a - 3)^2 = (a - 1)^2 + 2^2 \qquad\qquad (8.2)$$

Here b and a take the values of y and z, respectively, for $(z - 2)^2 + y^2 = z^2$, including both Plato and Pythagoras, so that I call this class the *PP-minus class*. This class includes $6^2 + 7^2 = 9^2 + 2^2$, $8^2 + 14^2 = 16^2 + 2^2$, $10^2 + 23^2 = 25^2 + 2^2$, and $12^2 + 34^2 = 36^2 + 2^2$. We can express the class in terms of Fibonacci's formula, with m an integer greater than 1, and $n = 1$, $p = 1$, and $q = m + 2$.

Method 2B: For squares in the PP-minus class.

1. Cut an x-square from the upper left corner of the z-square.

2. Cut a $(2 \times (m^2 - 1))$-rectangle from the lower left corner of the z-square.

3. Stutter-step starting 2 below the upper right corner of the z-square:
   $[\lfloor (m - 1)/2 \rfloor$ times]: {Move/cut left $2m$; Move/cut down 2.}
   Move/cut right $(m - 2)m - 1$; Move/cut down a distance of 2.
   $[\lfloor (m - 2)/2 \rfloor$ times]: {Move/cut left $2m$; Move/cut down 2.}
   Move/cut left $2(m - 1)$.

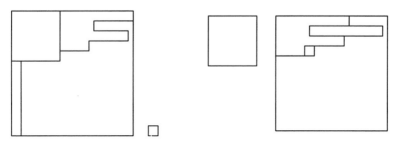

8.3: Method 2B applied to $10^2 + 23^2 = 25^2 + 2^2$

Note: We indicate rounding down to the nearest integer by using the symbols $\lfloor \ \rfloor$. Method 2B is illustrated in Figure 8.3 for $m = 4$, which corresponds to $10^2 + 23^2 = 25^2 + 2^2$, and in Figure 8.4 for $m = 5$. Notice how closely Figure 8.4 resembles Figure 7.5.

If the squares of sides x and y are attached, then we can modify Method 2B to give a 4-piece dissection. Suppose that the squares of sides z and w are also attached. Can we find a 3-piece dissection? Find it when m is odd:

Puzzle 8.1: Suppose that the squares of sides x and y are attached, and the squares of sides z and w are also attached. Adapt Method 2B to give a 3-piece dissection when m is odd.

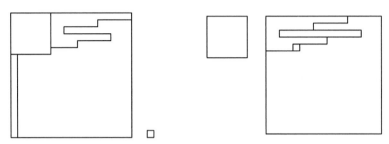

8.4: Method 2B applied to $12^2 + 34^2 = 36^2 + 2^2$

Our third class is rather different from the previous two, since it has two degrees of freedom rather than one in substituting into Fibonacci's formula. Take $m = 2$ and $n = 1$, and any values for p and q. The sums $x^2 + y^2$ and $z^2 + w^2$ then simplify to $5(p^2 + q^2)$. Hence I call this class of solutions the *Penta class*. It includes $5^2 + 5^2 = 7^2 + 1^2$, $7^2 + 4^2 = 8^2 + 1^2$, and $6^2 + 7^2 = 9^2 + 2^2$.

Method 3: For squares in the Penta class.

1. Cut a y-square from the upper left corner of the z-square.

2. Step starting q below the upper right corner of the z-square:
   Move/cut left p; Move/cut down p; Move/cut left p.
   If $2p - q > 0$, move/cut up $2p - q$.

3. Step starting $q - p$ to the right of lower left corner of the z-square:
   Move/cut up p; Move/cut right p; Move/cut up p.

Method 3 has several nice properties. First, it is translational, so that it works even for rhombuses whose side lengths satisfy the required relationship. Second, the w-square is always in the center in the x-square. Third, the pieces in the lower left and the upper right of the z-square fit together to give a copy of the piece in the lower right of the z-square.

Method 3 is illustrated in Figure 8.5 for $p = 1$ and $q = 4$, which corresponds to $6^2 + 7^2 = 9^2 + 2^2$. Since we interchange the position of the two step pieces in forming

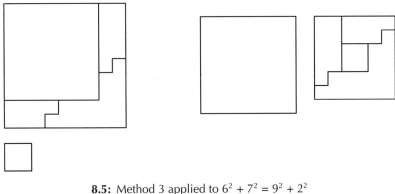

8.5: Method 3 applied to $6^2 + 7^2 = 9^2 + 2^2$

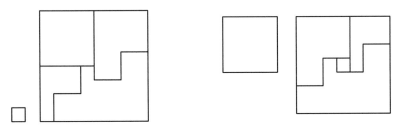

8.6: Method 3 applied to $7^2 + 4^2 = 8^2 + 1^2$

one square from the other, we shall call this the *interchanged step* technique. The method is also illustrated in Figure 8.6 for $p = 2$ and $q = 3$, which corresponds to $7^2 + 4^2 = 8^2 + 1^2$. When the x- and y-squares are attached, and the z- and w-squares are, too, then we get 3-piece dissections.

Puzzle 8.2: What dissection results from Method 3 when $2p - q = 0$?

Puzzle 8.3: Must p and q be rational numbers for Method 3 to work? What happens when $p = \sqrt{2}$ and $q = \sqrt{3}$?

The interchanged step technique has a wonderful double-duty application. Robert Reid suggested the puzzle of a three-way dissection involving pairs of squares: $x^2 + y^2 = z^2 + w^2 = v^2 + u^2$. I have discovered a class of solutions to these equations for which the associated dissection uses only eight pieces. To generate the class, take p and q to be relatively prime positive integers such that $q/2 < p < q$. Using these values, produce the solution $a^2 + b^2 = c^2 + d^2$ in the Penta class. Next use a and b in the roles of q and p, respectively, and generate another solution $z^2 + w^2 = v^2 + u^2$ in the Penta class. Then take c and d in the roles of q and p, respectively, and generate a third solution in the Penta class. This latter solution will share a term equal to $z^2 + w^2$, so that we can represent this last solution by $w^2 + z^2 = x^2 + y^2$. In terms of the original values of p and q, $x = 3q + 4p$, $y = 4q - 3p$, $z = 5q$, $w = 5p$, $v = 4q + 3p$, and $u = |3q - 4p|$. When we apply Method 3 to each of the solutions, we cut the w-square into four pieces in one dissection and the z-square into four pieces in the other. Overlaying these cuts produces no additional cuts, so that eight pieces suffice.

Since we use the Penta approach on two levels, we shall call the resulting class the *Penta–penta* class. Members of the class include $17^2 + 6^2 = 15^2 + 10^2 = 18^2 + 1^2$ (for $q = 3$, $p = 2$), $27^2 + 11^2 = 25^2 + 15^2 = 29^2 + 3^2$ (for $q = 5$, $p = 3$), and $31^2 + 8^2 = 25^2 + 20^2 = 32^2 + 1^2$ (for $q = 5$, $p = 4$). The dissection for the first solution is given in Figure 8.7, with the dissected 15-square in the upper left of the 17-square, and the dissected 10-square in the upper left of the 18-square.

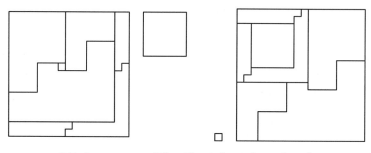

8.7: Penta–penta: $17^2 + 6^2 = 15^2 + 10^2 = 18^2 + 1^2$

Not satisfied with the selection of 5-piece dissections for $x^2 + y^2 = z^2 + w^2$, Robert Reid sharpened his chisel and found more classes. The next includes $14^2 + 12^2 = 18^2 + 4^2$, $23^2 + 24^2 = 32^2 + 9^2$, and $34^2 + 40^2 = 50^2 + 16^2$. It results from taking $n = 1$, $p = m$, and $q = 2m - 1$ in Fibonacci's formula. When p is odd, each term in the solution has a common factor of 2. We call this class the *square-difference class*, since $z - x$ is a square and equal to y. Reid identified half the solutions in this class and discovered Method 4. I then found the other half of the class, for which Reid's approach also works. Method 4 is illustrated in Figure 8.8 for $m = 4$, which corresponds to $23^2 + 24^2 = 32^2 + 9^2$.

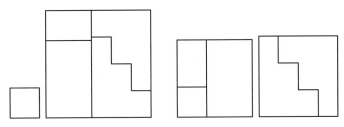

8.8: Method 4 applied to $23^2 + 24^2 = 32^2 + 9^2$

Method 4:   For squares in the square-difference class.

1. Cut an $((x - w) \times w)$-rectangle out of the upper left of the z-square.

2. Cut an $((x - w) \times x)$-rectangle from the lower left of the z-square.

3. Use the step technique to cut the remainder of the z-square,
giving $m - 1$ steps of height $q + 1$ and tread $q - 1$.

Another class of solutions that Reid found includes $13^2 + 18^2 = 22^2 + 3^2$, $22^2 + 32^2 = 38^2 + 8^2$, and $16^2 + 18^2 = 24^2 + 2^2$ and results from taking $n = 1$, $m = p$ or $m = p + 2$, and $q = 2p + 1$ in Fibonacci's formula. When p is odd, each term has a common factor of 2. Since $z - x$ is either a square or one less than a square, and y is twice this same square, I call this class the *double-square-difference class*.

Method 5:   For squares in the double-square-difference class.

1. Cut an x-square from the lower right corner of the z-square.

2. Start $2p$ to the right of the upper left of the z-square:
Move/cut down $y - q - 1$; Move/cut right $y - x$.

3. Step starting $2p(p + 1 - m)$ up from the lower left of the z-square if $m = p$, and from the upper left otherwise:
Move/cut right 2.
[$p - 1$ times]: {Move/cut up $2p(p + 1 - m)$; Move/cut right 2.}

Method 5 is illustrated in Figure 8.9 for $m = p + 2$ with $p = 2$, which corresponds to $13^2 + 18^2 = 22^2 + 3^2$. In this case, each step goes up by a negative amount; that means it actually goes down.

"What's that?... You say that you have found these designs by Robert Reid won- derfully distracting, but now you are impatient to see our world-famous lines of

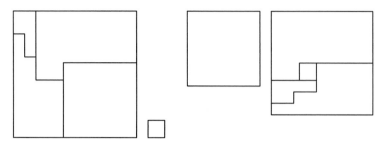

8.9: Method 5 applied to $13^2 + 18^2 = 22^2 + 3^2$

$x^2 + y^2 + z^2 = w^2$-cabinetry? Of course you can inspect the samples in our showroom warehouse! But first let's take a few minutes to go over some paperwork":

Finding integral solutions to $x^2 + y^2 + z^2 = w^2$ dates back to the time of the ancient Greeks, when Diophantus noted that $2^2 + 3^2 + 6^2 = 7^2$. We can find all integral solutions to $x^2 + y^2 + z^2 = w^2$ by using a method by the nineteenth-century French mathematician V. A. Lebesgue. Choose m, n, p, q to be positive integers with $m^2 + n^2 > p^2 + q^2$ and $mq > np$. Then *Lebesgue's formula* is

$$x = m^2 + n^2 - p^2 - q^2, \qquad y = 2(mp + nq), \qquad z = 2(mq - np),$$
$$w = m^2 + n^2 + p^2 + q^2$$

Interestingly, when $m = n$ and $q = p$, Lebesgue's formula reduces to twice Diophantus's formula (for generating solutions to $x^2 + y^2 = z^2$).

Besides Loyd's dissection discussed at the beginning of the chapter, we can find several dissections of squares for $x^2 + y^2 + z^2 = w^2$ in early puzzles. Dudeney (1907) gave several 4-piece dissections for $1^2 + 2^2 + 2^2 = 3^2$. Loyd (*Inquirer*, 1900a) gave a 3-piece dissection for $1^2 + 4^2 + 8^2 = 9^2$ if the three squares of sides 1, 4, and 8 are attached; it becomes a 6-piece dissection if those squares are not attached. Geoffrey Mott-Smith (1946) gave a 5-piece dissection for the same relation, which becomes a 3-piece dissection if the squares are attached, though the attachment is different from that in the Loyd dissection. The unattached version is shown in Figure 8.10, with the squares positioned to suggest the solution when the squares are attached. Dudeney (*Strand*, 1923a) gave a 6-piece dissection for $12^2 + 15^2 + 16^2 = 25^2$. Dudeney most likely derived the relation by noting that $15^2 + 20^2 = 25^2$ and $12^2 + 16^2 = 20^2$, thus using two multiples of $3^2 + 4^2 = 5^2$.

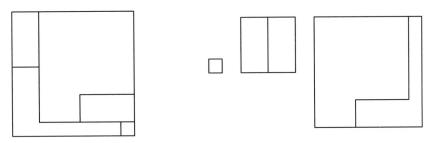

8.10: Mott-Smith's squares for $1^2 + 4^2 + 8^2 = 9^2$

Puzzle 8.4: Find a 4-piece dissection for $1^2 + 2^2 + 2^2 = 3^2$
that is translational.

More discerning aficionados will notice that my first two classes again share similarities with the Pythagoras and Plato classes from Chapter 7. For the first class, start with $x^2 + (z-1)^2 = z^2$, which gives us solutions in the class of Pythagoras. Take $z = a - 2$ and $x = b$, and then substitute for $(a-2)^2$ in equation (8.1), giving $b^2 + (a-1)^2 + 2^2 = a^2$. Since b and a take the values of x and $z + 2$, respectively, in the class of Pythagoras, I call this class the *Pythagoras-plus class*. This class includes $3^2 + 6^2 + 2^2 = 7^2$ and $5^2 + 14^2 + 2^2 = 15^2$. We can express it in terms of

---

Henry Ernest Dudeney was born at Mayfield, Sussex, England, in 1857 and died at Lewes in 1930. A self-educated mathematician who never attended college, Dudeney started in the civil service as a clerk at age thirteen. In his twenties, he began publishing short stories in the press and soon gave up his work as a clerk. He also began to publish puzzles under the pseudonym "Sphinx," wrote the chess column in *Wit and Wisdom*, and was a church organist. His puzzle columns appeared in a variety of periodicals for over thirty years. Collections of Dudeney's puzzles appeared in five books, two of which were published posthumously. He was in this regard less prodigious than his wife, Alice, who, as the author of over 20 popular novels, was more famous.

Dudeney maintained a lifelong interest in music. He enjoyed attending opera and playing piano transcriptions of Richard Wagner's works. He was equally fond of games, including billiards and croquet, in his youth. As an old man, until his last illness, he played bowls every evening on the bowling green within the picturesque Castle Precincts, Lewes, where he lived for the last ten years of his life.

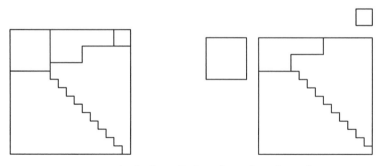

8.11: Method 1C applied to $5^2 + 14^2 + 2^2 = 15^2$

Lebesgue's formula, by taking $n = 1$, $q = 1$, and $m = p + 1$. This class satisfies relationship (4.4). Method 1C gives a general method for this class that corresponds to converting the P-slides in Figure 4.21 into steps.

Method 1C is illustrated in Figure 8.11 for the value $p = 2$, which corresponds to $5^2 + 14^2 + 2^2 = 15^2$. If the squares of sides x, y, and z are attached, then there are 3-piece dissections. The dissections are translational.

Method 1C: For squares in the Pythagoras-plus class.

1. Cut an x-square from the upper left corner of the w-square.

2. Cut a z-square from the upper right corner of the w-square.

3. Step starting 1 left of the lower right corner of the w-square:
  Move/cut up a distance of 1.
  [$w - x - 1$ times]: {Move/cut left 1; Move/cut up 1.}

4. Step starting at the lower left corner of the z-square:
  Move/cut left a distance of $2p$.
  [$p - 1$ times]: {Move/cut down 2; Move/cut left $2p$.}

The second class of solutions is derived from $(z - 2)^2 + y^2 = z^2$, which includes all solutions in Plato's class, as well as twice each solution in Pythagoras's class. Take $z = a - 1$ and $y = b$ and then substitute for $(a-1)^2$ in equation (8.1), giving

$$(a - 2)^2 + 2^2 + b^2 = a^2 \tag{8.3}$$

Since b and a take the values of y and $z + 1$, respectively, in the class of solutions to $(z - 2)^2 + y^2 = z^2$, I call this class the *PP-plus class*. This class includes $9^2 + 2^2 + 6^2 = 11^2$, $16^2 + 2^2 + 8^2 = 18^2$, and $25^2 + 2^2 + 10^2 = 27^2$. We can express it in terms of Lebesgue's formula, with $n = 1$, $q = 1$, and $p = 0$. This class satisfies relationship (4.3). Method 2C gives a general method for this class.

Method 2C: For squares in the PP-plus class.

1. Cut a z-square from the upper left corner of the w-square.

2. Cut a 2-square whose upper left touches the lower left of the z-square.

3. Cut a $(2 \times (m^2 - 2m))$-rectangle from the lower left of the w-square.

4. Stutter-step starting 2 below the upper right corner of the w-square:
    $[\lfloor (m-2)/2 \rfloor$ times]: {Move/cut left $2m-2$; Move/cut down 2.}
    Move/cut to the right a distance of $(m-3)(m-1) - 1$.
    $[\lfloor (m - 1)/2 \rfloor$ times]: {Move/cut down 2; Move/cut left $2m - 2$.}

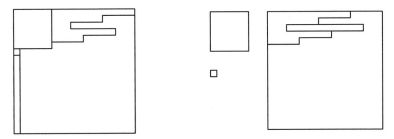

8.12: Method 2C applied to $36^2 + 2^2 + 12^2 = 38^2$

The method is illustrated in Figure 8.12 for the value $m = 6$, which corresponds to $36^2 + 2^2 + 12^2 = 38^2$. If the squares of sides x, y, and z are attached, then we can find 3-piece dissections. Although Method 2C produces a dissection in which one of the pieces is rotated, there are 5-piece translational dissections for $16^2 + 2^2 + 8^2 = 18^2$ and $8^2 + 1^2 + 4^2 = 9^2$.

Puzzle 8.5: Find a 5-piece translational dissection for $8^2 + 1^2 + 4^2 = 9^2$.

"What? You find the first two classes to be just cheap knock-offs of ancient designs?"

Well, the third class has a more original, yet still traditional feel. It was identified by the eighteenth-century Italian mathematician P. Cossali. It includes

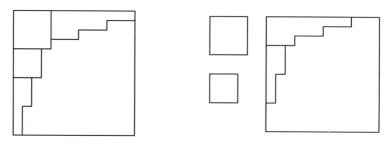

8.13: Squares for $3^2 + 12^2 + 4^2 = 13^2$

$3^2 + 6^2 + 2^2 = 7^2$, $3^2 + 12^2 + 4^2 = 13^2$, and $5^2 + 20^2 + 4^2 = 21^2$ and results from taking $m = p + 1$, $n = m$ or $n = p$, and $q = n$ in Lebesgue's formula. This class satisfies relationship (4.5) if we use $x' = \min\{x, z\}$ and $z' = \max\{x, z\}$. A general method for this class results from converting the P-slides in Figure 4.22 to steps.

The resulting dissection is translational and has reflection about the diagonal in the one large piece. The method is illustrated in Figure 8.13 for $n = 2$ and $p = 1$, which corresponds to $3^2 + 12^2 + 4^2 = 13^2$. If the squares of sides x, y, and z are attached, then there are 3-piece dissections.

The next class includes $1^2 + 8^2 + 4^2 = 9^2$, $9^2 + 8^2 + 12^2 = 17^2$, and $4^2 + 18^2 + 12^2 = 22^2$, and results from taking $p = 0$, $q = n$, and m such that $|m - n| = 1$ in Lebesgue's formula. In this class x is a square, y is twice a square, and $x + y = w$. Thus I call this the *square-sum class*. This class satisfies relationship (4.3). I discovered the set of solutions for $m - n = 1$, which I call the *square-sum-plus class*, and Robert Reid discovered those for $m - n = -1$, which I call the *square-sum-minus class*. A general method for this class results from converting the P-slide dissection for relationship (4.3) to a step.

The method is illustrated in Figure 8.14 for $n = 3$ and $m = 4$ and in Figure 8.15 for $n = 4$ and $m = 3$, which correspond to $16^2 + 18^2 + 24^2 = 34^2$ and $9^2 + 32^2 + 24^2 = 41^2$, respectively. The same step is used in both figures but is rotated $90°$ (and reflected) from one to the other. The method works whenever $x < y$. For $9^2 + 8^2 + 12^2 = 17^2$, corresponding to $n = 2$ and $m = 3$, we can use the dissection in Figure 8.14 since the dimensions are double those required. Although the method produces a dissection in which one of the pieces is rotated, there are (at least) two 5-piece dissections for $9^2 + 8^2 + 12^2 = 17^2$ that are translational.

Puzzle 8.6: Find a 5-piece dissection for $9^2 + 8^2 + 12^2 = 17^2$ that is translational.

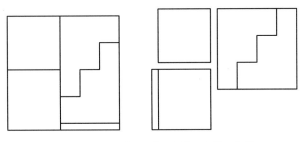

8.14: Squares for $16^2 + 18^2 + 24^2 = 34^2$

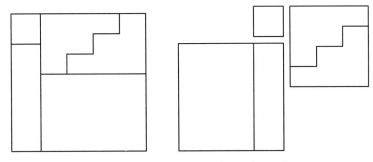

8.15: Squares for $9^2 + 32^2 + 24^2 = 41^2$

"If you aren't blurry-eyed yet, our final class will leave you seeing double!"

It is derived by adding the solutions of the PP-plus class to the corresponding solutions of the PP-minus class. Summing equations (8.2) and (8.3) gives $(a-3)^2 + b^2 + b^2 = (a+1)^2$, where b and a take the values of y and $z+1$, respectively, in the class of solutions to the Pythagorean equation with $x = z - 2$. The class includes $7^2 + 6^2 + 6^2 = 11^2$, $14^2 + 8^2 + 8^2 = 18^2$, $23^2 + 10^2 + 10^2 = 27^2$, and $34^2 + 12^2 + 12^2 = 38^2$. Since we combine the solutions to the PP-minus and PP-plus classes, and thus we apply equation (8.1) twice, we shall call this class the *PP-double class*. We can express the class in terms of Lebesgue's formula, with $p = 1$, $q = 1$, and $n = 0$. This class satisfies relationship (4.2). The general method for this class works when $m > 3$.

Method 6:   For squares in the PP-double class.

1. Cut a y-square from the upper left corner of the w-square.

2. Cut a z-square whose upper left touches the lower left of the y-square.

3. Cut a $(4 \times (w - y - z))$-rectangle from the lower left of the w-square.

4. Stutter-step starting 4 below the upper right corner of the w-square:

$[\lfloor(m-2)/2\rfloor$ times]: {Move/cut left $2m-4$; Move/cut down 4.}

Move/cut to the right a distance of $w-y-z-(2m-4)$.

$[\lfloor(m-1)/2\rfloor$ times]: {Move/cut down 4; Move/cut left $2m-4$.}

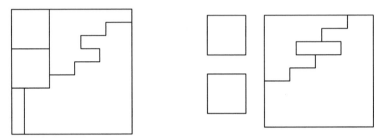

8.16: Method 6 applied to $34^2 + 12^2 + 12^2 = 38^2$

The method is illustrated in Figure 8.16 for the value $m = 6$, which corresponds to $34^2 + 12^2 + 12^2 = 38^2$. If the squares of sides x, y, and z are attached, then there are 3-piece dissections, as suggested by the positioning of the squares in the figure.

Method 6 does not apply if $m = 3$, since the sum of the side lengths of the y-square and the z-square exceeds the side length of the w-square. I found a 5-piece dissection for this case, but with one of the pieces turned over. Robert Reid has come to the rescue with a trim, attractive 5-piece dissection in which no pieces are turned over. His dissection for $7^2 + 6^2 + 6^2 = 11^2$ is given in Figure 8.17.

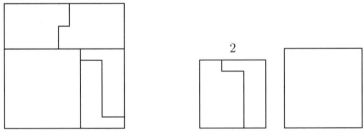

8.17: Reid's squares for $7^2 + 6^2 + 6^2 = 11^2$

"You find all these designs quite nice, really, but now you might prefer something that is a little bit more, well, one of a kind?"

An apparently isolated dissection is for $9^2 + 12^2 + 20^2 = 25^2$. We can derive this solution in a manner similar to that of Dudeney's $12^2 + 15^2 + 16^2 = 25^2$, but now using a different pair of multiples of $3^2 + 4^2 = 5^2$: $15^2 + 20^2 = 25^2$ and $9^2 + 12^2 = 15^2$. In this case, a 5-piece dissection is possible (Figure 8.18). The dissection is translational and uses only three pieces when the three squares are attached. This solution satisfies relationship (4.6). C. Dudley Langford noted that Ernest Irving Freese (1957) had a 5-piece dissection.

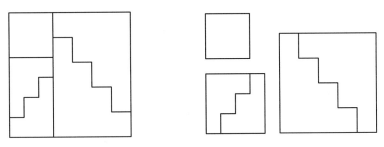

8.18: Squares for $9^2 + 12^2 + 20^2 = 25^2$

Another apparently isolated dissection is for $4^2 + 9^2 + 48^2 = 49^2$. The 5-piece dissection in Figure 8.19 has been discovered by Robert Reid. It is reminiscent of the one in Figure 8.13 but does not seem to be part of a larger class. Translational, it uses only three pieces when the three squares are attached.

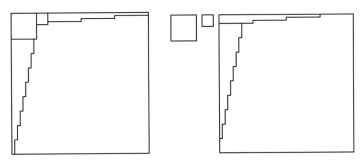

8.19: Reid's squares for $4^2 + 9^2 + 48^2 = 49^2$

"So you think now that three squares might not really be adequate, but that four squares could fit the bill?"

We do have a few examples of four squares to one. Freese (1957) gave the 6-piece dissection of squares for $2^2 + 4^2 + 5^2 + 6^2 = 9^2$, in Figure 8.20. The dissection is translational and uses only three pieces when the four squares are attached.

Dudeney (1917) posed another special case, for $1^2 + 4^2 + 8^2 + 12^2 = 15^2$, and gave a 4-piece dissection for the case when the squares are attached. A special set

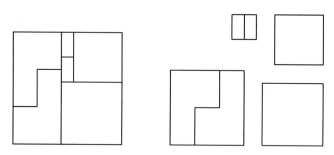

8.20: Freese's squares for $2^2 + 4^2 + 5^2 + 6^2 = 9^2$

of solutions to $x^2 + y^2 + z^2 + w^2 = v^2$ is $x = 1$, $y = z = n$, $w = n^2$, and $v = n^2 + 1$. There are 6-piece dissections for $n > 2$, and a 5-piece dissection for $n = 2$.

"We're temporarily out of stock of this last item, which we feel sure now is just the right one for you.... But we wouldn't dream of pressuring you into a decision. Why don't you think about it?":

Puzzle 8.7: Find economical dissections for
$1^2 + n^2 + n^2 + (n^2)^2 = (n^2 + 1)^2$.

CHAPTER 9

Watch Your Step!

Standing to one side of the soap box, the clown sports a greasepaint smirk on his face. On the box is printed the label Washington Soap – Free From Lye. *Could this be a satiric political cartoon from the closing years of the twentieth century? No, this is the illustration accompanying a Sam Loyd math puzzle from the opening years of that same century.*

Did you watch your step, you puzzled masses of the modern age? For you needed the same good measure of skepticism to solve that puzzle that our fed-up-and-not-taking-it-anymore citizenry now need to cut through all the splattered mud and political hocus-pocus. For Loyd had set the seemingly impossible task of dissecting a three-dimensional box to a two-dimensional square. If he himself had been on the square, he should have admitted that he intended that the projected outline of the box be dissected. And not to give away anything to modern trickery, a row of crooked steps was needed to solve the puzzle in the required two pieces!

The politic puzzling – or is it puzzle politicking? – that we indulge in here was prompted by the Jack and the Box Puzzle published by Loyd (*Press*, 1902b). We reproduce a facsimile of the box in Figure 9.1. To trip up his readers, Loyd used a modified form of the familiar step technique so that the required steps were not at right angles to each other. Deceptively simple, the 2-piece dissection of the box's outline to a square is shown in Figure 9.2. Perhaps there is another reason for the smirk on the clown's face, that this dissection was essentially this same one

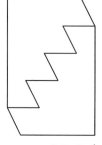
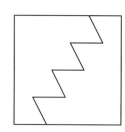

9.1: Box outline   **9.2:** Parhexagon to square

given already by Dudeney (*Dispatch*, 1899b). Does puzzle politics, like real politics, include appropriating another's good ideas?

In this chapter we take our cue from Loyd, and step away from past rectitude, by relaxing our notion of what a step should look like. This will allow us to apply the step technique to polygons other than squares. We will explore special cases of integral solutions to the equations $x^2 + y^2 = z^2$, $x^2 + y^2 = z^2 + w^2$, and $x^2 + y^2 + z^2 = w^2$ for figures such as triangles, pentagons, and hexagons. David Collison (1980) has already explored many of these puzzles and found a variety of ingenious approaches.

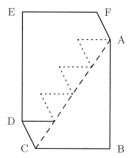
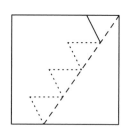

9.3: Conversion from H-slide

Before proceeding, we take a second look at Figure 9.2 and ask whether we can produce this dissection by converting a P-slide-like dissection, similar to that in Figure 7.2. A conversion from a new type of slide, which I call a *hexagon slide*, or *H-slide*, is shown in Figure 9.3. This slide takes a parhexagon, which is a hexagon in which opposite pairs of sides are parallel and equal, and converts it to a parallelogram. Specifically, it makes a cut from a vertex A and its next-but-one neighbor C, and then a second cut from the following neighbor D, with this cut parallel to the sides BC and EF. The resulting parallelogram has two sides parallel to AB and ED, and two more parallel to BC and EF. Although the H-slide appears not particularly useful at first, we shall see that it can explain part of Gavin Theobald's heptagon-to-square dissection in Figure 11.30.

  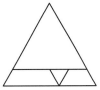

9.4: Bradley's triangles for $3^2 + 4^2 = 5^2$

We step gingerly at first, with dissections of two triangles to one. Harry Bradley (1930) described a 4-piece dissection of triangles for $3^2 + 4^2 = 5^2$. Shown in Figure 9.4, it leaves the 4-triangle uncut. This dissection still works if the triangles are not equilateral, but merely similar. Bradley also described a 4-piece dissection in which the 3-triangle is uncut.

Puzzle 9.1: Find a 4-piece dissection of triangles for $3^2 + 4^2 = 5^2$ in which the 3-triangle is uncut.

The existence of Bradley's dissection suggests that triangles in Pythagoras's class may also have 4-piece dissections. And David Collison identified such an approach (Method 7), which is illustrated in Figure 9.6 for the solution $5^2 + 12^2 = 13^2$, for which $n = 2$. Recall from Chapter 7 Diophantus's method for expressing basic solutions to $x^2 + y^2 = z^2$, which sets $x = m^2 - n^2$, $y = 2mn$, and $z = m^2 + n^2$, and remember that $m = n + 1$ for Pythagoras's class.

Method 7:   Collison's method for triangles in the class of Pythagoras.

1. Cut an x-triangle from the top of the z-triangle.

2. Cut an inverted n-triangle with right corner 1 to the left of the right corner of the x-triangle.

3. Step starting 1 right of the left corner of the z-triangle:
   [$2n - 1$ times]: {Move/cut up-right n; Move/cut right 1.}

The term *up-right* describes the direction up and to the right at an angle of $60°$. Similarly, *up-left* means up and to the left at an angle of $60°$, *dn-right* means

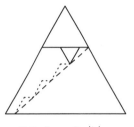
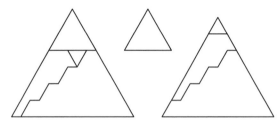

9.5: From P-slide   **9.6:** Collison's $5^2 + 12^2 = 13^2$

down and to the right at an angle of 60°, and *dn–left* means down and to the left at an angle of 60°.

Interestingly, this dissection is a conversion of Bradley's dissection (Figure 5.2). A step replaces the P-slide in the manner discussed in Chapter 7. We see an illustration of this conversion in Figure 9.5, with dashed lines indicating cuts that we eliminate, and dotted lines indicating cuts that we add. Since the inverted triangle is flipped up when the step piece is placed in the *y*-triangle, I call this the *flip-up step* technique.

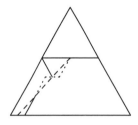
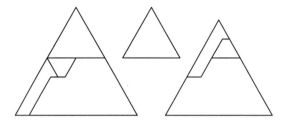

9.7: From Q-slide   **9.8:** Collison's $15^2 + 8^2 = 17^2$

Is there a method that gives 4-piece solutions for members of Plato's class? Yes, David Collison found a method for this class too, but it is actually a bit more general. In particular, it works for all solutions to $x^2 + y^2 = z^2$ for which $z = x + 2$. Since these consist of Plato's class plus the double of solutions in Pythagoras's class, I call this class the *PP class*. Collison's method is illustrated in Figure 9.8 for the solution $15^2 + 8^2 = 17^2$.

This method turns out to be a conversion of Lindgren's dissection of two triangles to one, shown in Figure 5.3. The conversion of the Q-slide into a modified step dissection is illustrated in Figure 9.7, with dashed lines indicating cuts that we eliminate, and dotted lines indicating cuts that we add. The last step up in Figure 9.7 is cut and flipped down when the step piece is positioned in the *y*-triangle. I thus call this technique the *flip-down step* technique.

Having flipped with triangles for $x^2 + y^2 = z^2$, can we slip with triangles for $x^2 + y^2 = z^2 + w^2$? I have found two methods that together handle a large number of these solutions. The first applies whenever $m = n + 1$ or $m = n + 2$. Since $m = n + 1$ is the restriction that gives Pythagoras' class in Diophanstus's formula for $x^2 + y^2 = z^2$, I call this class the *Pythagoras-extended* class. This class is a proper superset of the Pythagoras-minus, PP-minus, and Penta classes introduced in Chapter 8.

Method 8:   For triangles in the Pythagoras-extended class.

1. Cut a y-triangle from the top of the z-triangle.

2. Flip-down step starting $z - x$ right of the left corner of the z-triangle:
$$[\lfloor (n - 1)/(m - n) \rfloor \text{ times}]:$$
$$\{\text{Move/cut up-right } 2(m - n)p; \text{Move/cut right } z - x.\}$$
Move/cut up–right $2p$.

3. Move/cut dn–right all the way through the step piece,
starting $nq - mp$ right of the top left if $nq > mp$,
and otherwise starting $mp - nq$ dn–left of the top left.

In Figure 9.9, we illustrate Method 8 in the case of $m = n + 1$, with $n = 3$, $p = 1$, and $q = 2$, which gives $10^2 + 5^2 = 11^2 + 2^2$. In this case, $nq > mp$, so that we cut the step piece to the right of its upper left corner. Figure 9.10 illustrates $m = n + 2$, with $n = 3$, $p = 4$, and $q = 5$, which gives $35^2 + 13^2 = 37^2 + 5^2$. In this

David Michael Collison was born in Chelmsford, Essex, England, in 1936. A King's Scholar at Westminster School in London, he went on to Magdalene College at Cambridge University, where he earned a B.A. and an M.A. in mathematics, in 1959 and 1963, respectively. Entering the field of computing in its early years, he worked for several different companies in the United Kingdom before transferring to the United States. From 1965 to 1985 he worked at a succession of companies as senior programmer and eventually software engineer. From 1985 until his death in 1991 he was a freelance computer consultant.

Mathematical problems were a major focus of David's leisure. He spent many years researching and writing an as yet unpublished book on magic squares. He also read widely in nonmathematical disciplines, wrote poetry, and enjoyed music, chess, photography, cycling, and collecting antique radios.

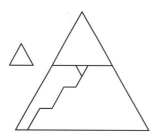
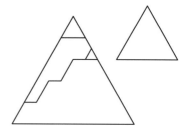

9.9: Method 8 applied to $10^2 + 5^2 = 11^2 + 2^2$

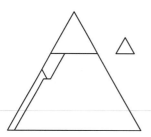
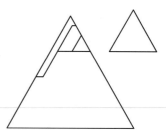

9.10: Method 8 applied to $35^2 + 13^2 = 37^2 + 5^2$

case, $nq < mp$, so that we cut the step piece to the down–right of its upper left corner.

In dissection puzzles as well as politics, it is handy to have an abundance of fancy footwork. My second method for $x^2 + y^2 = z^2 + w^2$ applies whenever $n = 1$. Since $n = 1$ is the restriction that gives Plato's class in Diophantus's formula for $x^2 + y^2 = z^2$, I call this class the *Plato-extended* class. This class is a proper super-set of the PP-minus, Penta, square-difference, and square-double-difference classes introduced in Chapter 8. We illustrate Method 9 with the example in Figure 9.11. For this case, $m = 4$, $p = 2$, and $q = 5$.

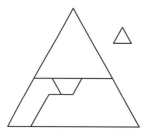
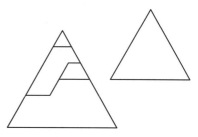

9.11: Method 9 applied to $13^2 + 18^2 = 22^2 + 3^2$

Method 9: For triangles in the Plato-extended class.

1. Cut an x-triangle from the top of the z-triangle.

2. Flip-down step starting $2p$ right of the left corner of the z-triangle:
    $[\hat{\Uparrow}(m-2)/2\degree \text{ times}]$:
        {Move/cut up-right $2(q-p)$; Move/cut right $2p$.}
    Move/cut up-right $2(q-p)$ if m is odd, and $q-p$ otherwise.

3. Move/cut dn–right all the way through the step piece,
    starting $mp-nq$ right of the top left.

So now you must feel "step-wise" and ready to handle the zigs and zags of triangles for $x^2 + y^2 + z^2 = w^2$. I know of economical dissections for three classes: Cossali's, the square-sum-plus, and the PP-double classes. David Collison found an elegant method for triangles in Cossali's class, which uses only four pieces. It converts the P-slide used in Figure 5.13 to a step.

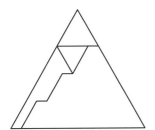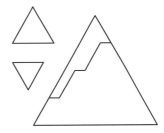

9.12: Collison's triangles for $3^2 + 4^2 + 12^2 = 13^2$

Collison's method is illustrated in Figure 9.12 for the solution $3^2 + 4^2 + 12^2 = 13^2$, for which $m + n = 4$. This is a conversion of three triangles to one when $x^2 = (w-z)(w-y)$, shown in Figure 5.13.

My dissection of three triangles to one in the square-sum-plus class is a conversion of the solution to Puzzle 5.3. We illustrate it in Figure 9.13 for $n = 2$, which produces $9^2 + 8^2 + 12^2 = 17^2$.

Method 10: For triangles in the PP-double class.

1. Cut a y-triangle from the right corner of the w-triangle.

2. Step starting at 4 up–right from the left corner of the w-triangle:
    $[\lceil m/2 \rceil - 1 \text{ times}]$: {Move/cut right $2m-4$; Move/cut up–right 4.}
    Move/cut right $m-2$ if m is odd,
        and right $2m-4$, then up–right 2, otherwise.

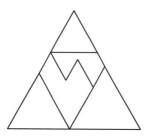

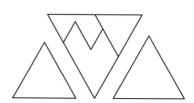

9.13: Triangles for $9^2 + 8^2 + 12^2 = 17^2$

3. Starting at the rightmost point of the step piece, move/cut left.

4. Starting at the lower right corner of the step piece,
    move/cut up–left $2m - 2$.

My method for the PP-double class is described by Method 10. This dissection also turns out to be another conversion, this time of Figure 5.10. We illustrate Method 10 in Figure 9.14 for $m = 3$, which corresponds to $7^2 + 6^2 + 6^2 = 11^2$.

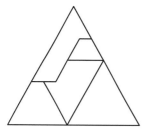

9.14: Method 10 applied to $7^2 + 6^2 + 6^2 = 11^2$

Recall from the last chapter the set of solutions to $x^2 + y^2 + z^2 + w^2 = v^2$ for which $x = 1$, $y = z = n$, $w = n^2$, and $v = n^2 + 1$. David Collison found 5-piece dissections for triangles in this class. A dissection is shown for $1^2 + 3^2 + 3^2 + 9^2 = 10^2$ in Figure 9.15.

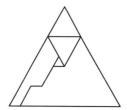

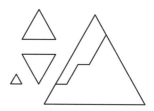

9.15: Collison's triangles for $1^2 + 3^2 + 3^2 + 9^2 = 10^2$

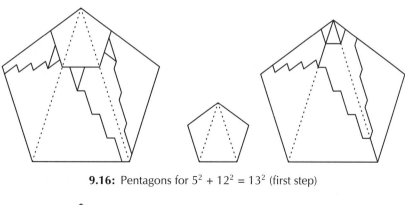

9.16: Pentagons for $5^2 + 12^2 = 13^2$ (first step)

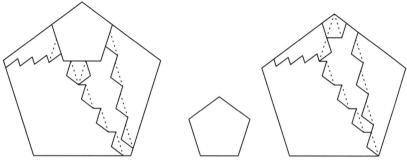

9.17: Collison's pentagons for $5^2 + 12^2 = 13^2$ (final)

Now that you see through our triangulation strategy, we will have David Collison maneuver in the "heavy armor," exemplified by the pentagon. For regular polygons with n sides, realizing solutions in Pythagoras's class, he found $(n + 1)$-piece dissections when n is odd, and n-piece dissections when n is even. He based these dissections on several nice ideas. First, he viewed an $\{n\}$ as a union of $n - 2$ triangles and applied the triangle dissection of Method 7 to each, giving a $4(n - 2)$-piece dissection. His second idea was to handle alternate triangles by the reflection of the standard triangle dissection and to merge two pairs of like pieces from each pair of neighboring triangle dissections, giving a $2(n - 1)$-piece dissection. Finally, he combined the unit triangles by adding the unit triangle to the first piece, cutting a gap in the second piece, and so on. This gave a $(n + 1)$-piece dissection when n is odd.

We illustrate Collison's approach for pentagons realizing $5^2 + 12^2 = 13^2$. The $2(n - 1)$-piece dissection is shown in Figure 9.16, with dotted edges indicating where pieces were combined. The $(n + 1)$-piece dissection is shown in Figure 9.17, with dotted edges indicating where portions were added to and cut off various pieces.

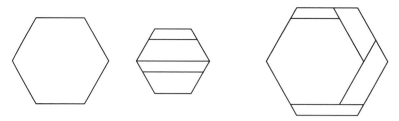

9.18: Schmerl's hexagons for $3^2 + 4^2 = 5^2$

Judging from the first example in this chapter, we might expect hexagons to be politically correct. So let's find out! James Schmerl (1973) gave a 5-piece dissection of hexagons for $3^2 + 4^2 = 5^2$, shown in Figure 9.18. In this dissection Schmerl left the 4-hexagon uncut. You can also get a 5-piece dissection in which the 3-hexagon is uncut and there is bilateral symmetry.

Puzzle 9.2: Find a 5-piece dissection of hexagons for $3^2 + 4^2 = 5^2$ in which the 3-hexagon is uncut and there is reflection symmetry.

Collison found 6-piece dissections for hexagons in Pythagoras's class. Figure 9.19 gives his dissection for $5^2 + 12^2 = 13^2$. Rather than exhaust my energy and your patience, I do not give a derivation of his hexagon method here. You can try it yourself. Once the step piece in the upper right is chosen, you can infer the shape of every other piece, proceeding from right to left.

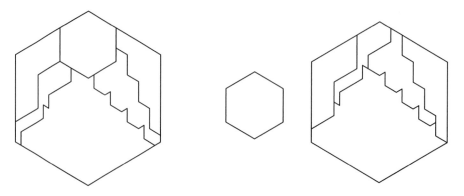

9.19: Collison's hexagons for $5^2 + 12^2 = 13^2$

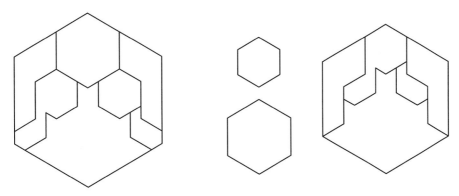

9.20: Collison's hexagons for $2^2 + 3^2 + 6^2 = 7^2$

James Schmerl has been a member of the math faculty at the University of Connecticut since 1972. His specialty is logic, in particular model theory, but he has enjoyed working occasionally in other areas. One of his recent papers investigates tiling space with translated copies of a notched cube.

Do any of the relationships for three squares to one carry over to economical dissections of three hexagons to one? The answer is a resounding yes, reflecting two beautiful methods described by David Collison. Both apply to Cossali's class, and we reproduce the dissection for $2^2 + 3^2 + 6^2 = 7^2$ in Figure 9.20.

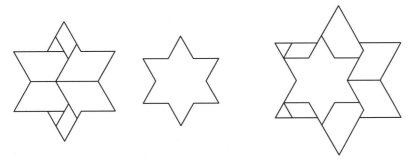

9.21: Hexagrams for $3^2 + 4^2 = 5^2$

Since we fared well with hexagrams in Chapter 5, let's try them again here. I give my 9-piece dissection for $3^2 + 4^2 = 5^2$ in Figure 9.21. Just as James Schmerl's dissection of hexagons takes special advantage of the dimensions 3-4-5, this dissection also takes advantage of these dimensions in a way that does not seem

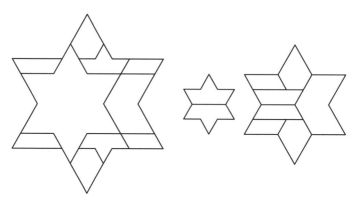

9.22: Reid's hexagrams for $1^2 + 2^2 + 2^2 = 3^2$

to generalize to other solutions of $x^2 + y^2 = z^2$. David Collison gave a beautiful method for stars of the form $\{2n/(n-1)\}$ in Pythagoras's class. However, his use of the $4n + 1$ pieces is no improvement on my dissection in Figure 5.24, which works for a much larger class of solutions.

Robert Reid has found a nifty 10-piece dissection of hexagrams for $1^2 + 2^2 + 2^2 = 3^2$. Shown in Figure 9.22, it is translational and has a pleasing bilateral symmetry.

We straighten ourselves out, at least temporarily, with my final set of rational dissections, involving the Greek Cross. I found that a simple approach gives 10-piece dissections for all members of Plato's and Pythagoras's classes except $3^2 + 4^2 = 5^2$. Cut the right arm off each Greek Cross and nestle it in the cavity

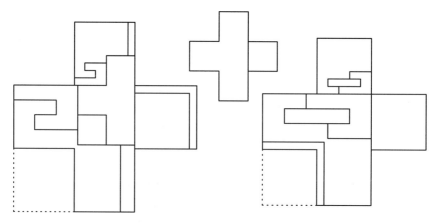

9.23: Greek Crosses for $15^2 + 8^2 = 17^2$

below the left arm. The resulting figure corresponds to two attached squares, with one twice the size of the other. Then apply either the Collison-like approach suggested by Figure 7.8 for Pythagoras's class or Method 2A for Plato's class. We illustrate the approach in Figure 9.23 for the case of $15^2 + 8^2 = 17^2$, with dotted lines showing where the right arm of the cross would be nestled in the 17-cross and the 15-cross.

This approach does not give a 10-piece dissection for $3^2 + 4^2 = 5^2$, because twice 3 is not smaller than 5, so that some cuts will overlap. But Robert Reid has found the clever 7-piece dissection in Figure 9.24.

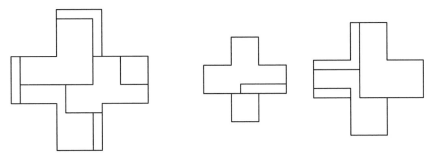

9.24: Reid's Greek Crosses for $3^2 + 4^2 = 5^2$

The case of $1^2 + 2^2 + 2^2 = 3^2$ also turns out to be interesting for Greek Crosses. At least two of the small Greek Crosses must be cut. In Figure 9.25 I give a 7-piece dissection that leaves a 2-Greek Cross uncut.

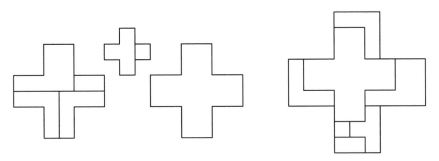

9.25: Greek Crosses for $1^2 + 2^2 + 2^2 = 3^2$

Puzzle 9.3: Find a 7-piece dissection of Greek Crosses for $1^2 + 2^2 + 2^2 = 3^2$, such that the 1-Greek Cross is uncut.

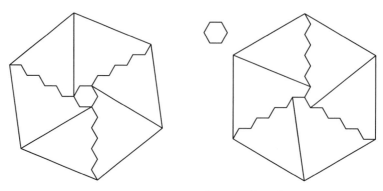

9.26: Hexagons for $1^2 + (\sqrt{48})^2 = 7^2$

Have the Greek Cross dissections left us drawn to the straight and narrow in the end? No, for we now make an enormous "misstep": We conclude this chapter with several dissections that we shall call *semirational*. By this we mean that some of the cuts are of a length that is a rational fraction of a side length and others are not. These dissections are all for two hexagons to one. Since we know of general dissections for two hexagons to one that have no more than eight pieces, we focus on special cases for which the number of pieces is at most seven and identify three classes.

The first class consists of solutions of $x^2 + y^2 = z^2$ for which $x = 1$, $y = \sqrt{3n^2}$, and $z = \sqrt{3n^2 + 1}$, which I call the *Tri-root* class. For $n = 1$, the solution to Problem 5.5 gives a 6-piece dissection. For $n > 1$ I have found a method that gives 7-piece dissections, illustrated for $n = 4$ in Figure 9.26. Besides having 3-fold rotational symmetry and translation with no rotation, the pieces from the y-hexagon can be cyclicly hinged, as shown in Figure 3.14.

Method 11: For hexagons in the Tri-root class.

Orient the y-hexagon with two sides vertical.

Repeat 3 times:

> Step from the top vertex:
>
>> Move/cut dn–right 1.
>>
>> [$n - 1$ times]: {Move/cut dn–left 1; Move/cut dn–right 1.}
>>
>> Mark the current position.
>>
>> Move/cut dn–left 1.
>
> Move/cut from the last mark to the upper left vertex.
>
> Rotate 120° clockwise around the center.

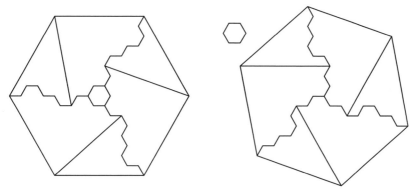

9.27: Hexagons for $1^2 + (\sqrt{63})^2 = 8^2$

The second class consists of $x^2 + y^2 = z^2$ for which $x = 1$ and $z = 3n^2 - 3n + 2$, which I call the *Tri-minus* class. When $n = 1$, I have the same solution as for $n = 1$ in the previous class. For $n > 1$, I have found a 7-piece dissection, illustrated for $n = 2$ in Figure 9.27. The steps in this method are less regular than for other dissections in this chapter, so I call it the *wobbly step* technique. Again, the dissection has 3-fold rotational symmetry, translation with no rotation, and the property that the pieces from the y-hexagon can be cyclicly hinged.

Method 12: Wobbly steps for hexagons in the Tri-minus class.

Orient the z-hexagon with two sides horizontal.

Cut a 1-hexagon with two sides horizontal from the center.

Repeat 3 times:

    Wobbly step from the left vertex:

        Move/cut right 1.

        $[n - 1$ times]:

            $\{ [n - 1$ times]: {Move/cut up–right 1; Move/cut right 1.}

            $[n$ times]: {Move/cut dn–right 1; Move/cut right 1.} $\}$

        Mark the current position.

        $[n - 1$ times]: {Move/cut up–right 1; Move/cut right 1.}

    Move/cut from the last mark to the upper left vertex.

    Rotate 120° clockwise around the center.

The third class consists of $x^2 + y^2 = z^2$ for which $x = 1$ and $y = 3n(n + 1)$, which I call the *Tri-plus* class. For $n \geq 1$ I have found a 7-piece dissection, once

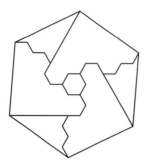 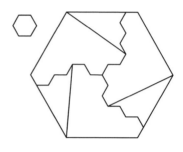

9.28: Hexagons for $1^2 + 6^2 = (\sqrt{37})^2$

again using the wobbly step technique. The method is illustrated for the case when $n = 1$ in Figure 9.28. Although the dissection has 3-fold rotational symmetry and translation with no rotation, this time its pieces cannot be hinged.

Method 13:   For hexagons in the Tri-plus class.

    Orient the y-hexagon with two sides horizontal.

    Repeat 3 times:

        Wobbly step starting 1 dn–right from the left vertex:

            [n times]:

                { [$n - 1$ times]: {Move/cut right 1; Move/cut dn–right 1.}

                [$n + 1$ times]: {Move/cut right 1; Move/cut up–right 1.} }

        Mark the current position.

        [n times]: {Move/cut right 1; Move/cut dn–right 1.}

        Move/cut right 1.

        Move/cut from the last mark to the lower left corner.

    Rotate 120° clockwise around the center.

CHAPTER 10

Just Tessellating

*At long last the train pulled out of Charing Cross sta-
tion and away from the fogbound throng of idolizers
and skeptics. The couple on board would share just a
few more days together before their two lives, so dis-
tinctively patterned and strikingly overlaid, were forever
cut asunder. Eli Lemon Sheldon, an American mortgage
banker posted in London, who under the pseudonym
"Don Lemon" had engaged literally Everybody in the
1890s with his six-penny anthologies, would be dead
within a year. His wife, Mary French-Sheldon, resident of
New York and daughter of a spiritualist and faith healer,
would soon, under the appellation "Bébé Bwana" (or
"Woman Master"), lead an expedition through East Africa
and then return to write a popular book and produce her
own World's Fair exhibit based on her adventures. Had
this superposition of unusual lives catalyzed in some in-
explicable manner the one remarkable dissection found
in one of Lemon's books?*

The dissection referred to here is that of a Greek Cross to a square in four pieces,
which is found in (Lemon 1890) and reproduced in Figure 10.2. And although Eli
Lemon Sheldon lived an unusual life, it appears that this dissection should not be
attributed to him. Jerry Slocum (1996) reproduces a $2\frac{1}{2}$ inch $\times$ $4\frac{1}{2}$ inch advertising
card for Scourene, a scouring soap manufactured by the Simonds Soap Company
of New York City. The puzzle on the card shows a picture of a Greek Cross and
asks for two straight cuts that divide the cross into pieces that form a square. On

the basis of the style of the card, Jerry Slocum attributes its design to Sam Loyd. The card gave a deadline of March 31, 1887, for submitting a solution to win a prize.

This dissection can be derived by superposing a tessellation of Greek Crosses with a tessellation of squares, as shown in Figure 10.1. Infinitely many four-piece dissections result from shifting one tessellation relative to the other. A symmetrical dissection can be obtained by positioning the corners of the squares in the centers of the crosses. Since the dissection in (Lemon 1890) is not symmetrical, it was probably not discovered by superposing tessellations. Instead, it may have been discovered by adapting the 9-piece dissection of five squares to one square given by Abū'l-Wafā's method in Chapter 6. In any event, the pieces in the Greek Cross can be variously fully hinged, though they then assemble differently to form the square.

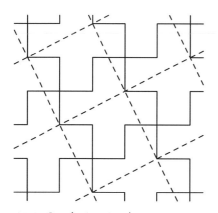

10.1: Overlaying Greek Cross, square

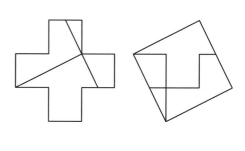

10.2: Greek Cross to square

"Don Lemon" was the pseudonym of **Eli Lemon Sheldon**, a banker who was born in Michigan in 1848. He was the European manager of the Jarvis-Conklin Mortgage Company, of America and London, from the time of its incorporation until his death in London in 1892. Sheldon was the editor of a number of compact books of reference such as *Everybody's Pocket Cyclopaedia of Things Worth Knowing*, *Everybody's Writing-Desk Book*, and *Everybody's Book of Short Poems*. He and his wife, Mary French-Sheldon, translated from French a book on Japan by Félix Régamey. He bankrolled, in the amount of $50,000, his wife's journey with a caravan of 138 men from Zanzibar to Kilima Njaro and back. This remarkable journey is chronicled in her book *From Sultan to Sultan*.

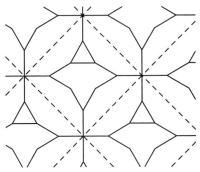

10.3: Tessellations: {12}, {4}

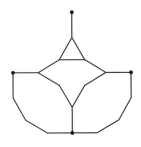

10.4: Dodecagon element

One of the most attractive dissections that arise from superposing tessellations is a dodecagon to a square, discovered by Lindgren (1951). He first cut the dodecagon into four pieces that rearrange to give a tessellation element in Figure 10.4. He then superposed the associated tessellation over a tessellation of squares (Figure 10.3) to give a 6-piece dissection (Figure 10.5).

We know from Figures 1.2 and 4.3 that there are simple dissections of a dodecagon to three equal squares, and of a square to two equal squares. Thus there should be an economical dissection of the dodecagon to $3 * 2 = 6$ squares, or alternatively, a Latin Cross. Lindgren (1962c) also identified elements for the dodecagon and Latin Cross whose tessellations give a neat 7-piece dissection. He cut the dodecagon into three pieces and rearranged them as in Figure 10.7, with the small circles representing the repetition points. He then cut the Latin Cross into two pieces and rearranged them as in Figure 10.8. Superposing the tessellations in Figure 10.6 gives the dissection in Figure 10.9.

Lindgren (1951) discovered the following curious feature: The set of pieces that forms the dodecagon element in Figure 10.7 also forms a second element, which has

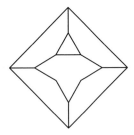

10.5: Lindgren's dodecagon to square

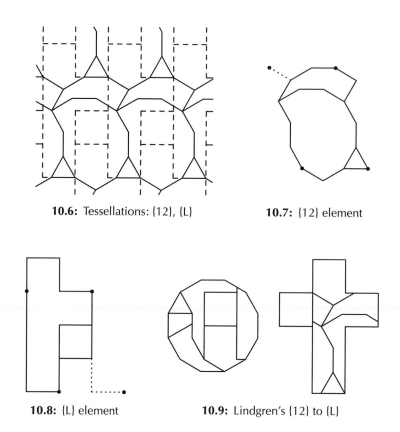

10.6: Tessellations: {12}, {L}

10.7: {12} element

10.8: {L} element

10.9: Lindgren's {12} to {L}

the repetition pattern of a square. This element then leads to a 6-piece dissection of a dodecagon to a square.

Puzzle 10.1: Rearrange the pieces in Figure 10.7 to form a tessellation element with the repetition pattern of a square. Find a corresponding dissection of a dodecagon to a square.

Lindgren's success in the last dissection depended on a substantial effort to convert each of the figures to tessellations with the same repetition patterns. In our next dissection, all of the work is in converting one of the two figures. We start with a 10-piece dissection of {12/3} to a square in (Frederickson 1972d), then cut away four outside pieces of the {12/3}, leaving a square as shown in Figure 10.10. The four pieces form a figure with the repetition pattern of a square, so we can dissect them to a square and then combine it with the larger square. But it is easier

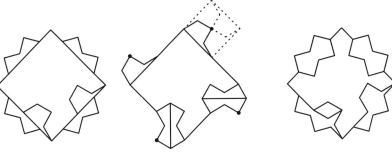

10.10: First {12/3} element   **10.11:** Second cut

to cut two notches in the larger square and rearrange the seven pieces to form a figure with the repetition pattern of the desired square.

It is possible to save a piece by combining with adjacent pieces the pieces resulting from cutting notches and then cutting notches for these combined pieces! The partition of the {12/3} is shown in Figure 10.11, and the new element is shown in Figure 10.12. Two of the pieces are turned over in the resulting 9-piece dissection (Figure 10.13).

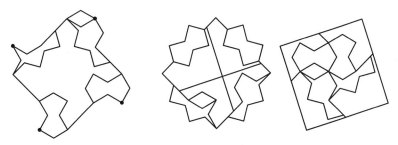

10.12: Second element   **10.13:** {12/3} to square

Lindgren (1964b) discovered how to cut the {8/2} into five pieces that form a tessellation element, which led to an 8-piece dissection of an {8/2} to a square. However, I improved this to seven pieces in (Frederickson 1972d) and give a variation of it in Figure 10.15. In Figure 10.14, we cut the star into four pieces to form a tessellation element and superpose its tessellation with one for squares in Figure 10.16.

We will discuss in Chapter 16 how this dissection is made possible by a general relationship between $\{(2n+4)/2\}$ and $\{n+2\}$ for each positive integer n. As we shall see, at most five pieces are needed for {6/2} to {3}, seven pieces for {10/2} to {5}, and eight pieces for {12/2} to {6}. Can a 6-piece dissection be found for the {8/2} and square? I have found a number of different 7-piece dissections.

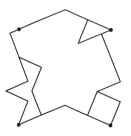
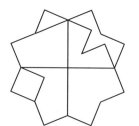
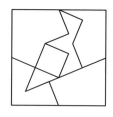

10.14: {8/2} element   **10.15:** {8/2} to square

Puzzle 10.2: Find a 7-piece dissection of an {8/2} to a square in which one piece has one full side of the square plus a half of another side.

Lindgren (1964b) also gave a 10-piece dissection of a {12/2} to a hexagon, which again inspired me. I found an 8-piece solution in (Frederickson 1972d) and give a variation of it in Figure 10.19. In Figure 10.18 we cut the {12/2} star into five pieces that form a tessellation element, with the repetition pattern of a hexagon. The superposition of that tessellation with one for hexagons, shown in Figure 10.17, gives the dissection.

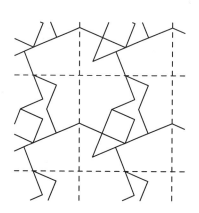
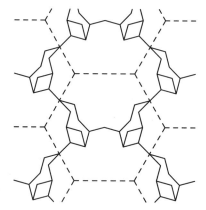

10.16: Overlay {8/2}, square   **10.17:** Overlay {12/2}, hexagon

After the last two examples, it is not surprising that there is an elegant dissection of a hexagram to a triangle. Geoffrey Mott-Smith (1946) found several 5-piece dissections, of which we see one in Figure 10.21. This dissection can be fully and variously hinged. Lindgren (1961) gave another 5-piece dissection, completely different.

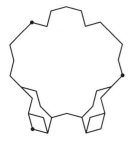 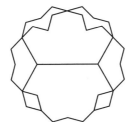 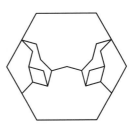

10.18: {12/2} element   **10.19:** {12/2} to hexagon

Mott-Smith's dissection can be discovered by using tessellations, with an element for the hexagram shown in Figure 10.20. Dashed edges indicate a tessellation element for a triangle. We match edges with short dashes together, and similarly edges with longer dashes, but the latter must also match an out-arrow label with an in-arrow label. We join repetition points only if they have the same label, A or B. Figure 10.22 shows a portion of a tessellation. If we add two more hexagram

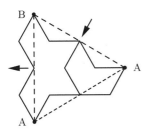 

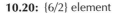

10.20: {6/2} element   **10.21:** Mott-Smith's {6/2} to triangle

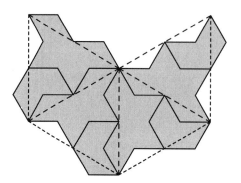

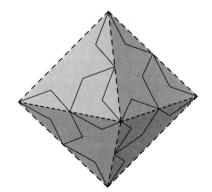

10.22: Tessellation for {6/2}

10.23: On an octahedron

elements, then we form a larger element with a hexagonal repetition pattern. Alternatively, we can fold the four hexagram elements on the dashed lines to give what will cover the top half of an octahedron. Thus eight of our hexagram elements tile an octahedron, as shown in Figure 10.23.

Dissecting a {10/2} to a {5} at first appears difficult, since we cannot tessellate the plane with pentagons. Lindgren (1964b) used trial and error to find an 8-piece dissection. However, we can use the notion of tessellations on the surfaces of Platonic solids, as suggested by Figures 10.20 and 10.23. I gave a 7-piece dissection in (Frederickson 1974).

A *polyhedral tessellation* is a tiling of the faces of a regular polyhedron by an element. We use a dodecahedral tessellation with the element shown in Figure 10.24 to create the dissection in Figure 10.25. The repetition points are the vertices of a pentagon, and the sides of the pentagon are indicated with dashed edges. The full dodecahedral tessellation is shown in Figure 10.26. Folding the twelve corresponding elements along the dashed edges and matching them up yield the dodecahedron shown in Figure 10.27. If we reexamine Figures 6.18 and 6.19, we will recognize elements for two other dodecahedral tessellations.

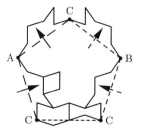

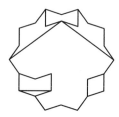

10.24: {10/2} element

10.25: {10/2} to pentagon

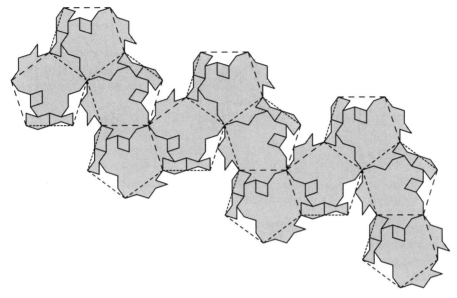

10.26: Dodecahedral tessellation for {10/2}

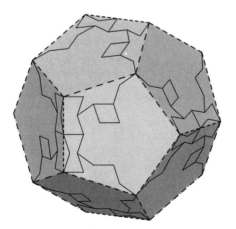

10.27: On a dodecahedron

Having started this chapter with a dissection of Greek Cross to square, we now turn to the Cross of Lorraine {L'}, constructed from 13 small squares. Since $13 = 2^2 + 3^2$, we expect a nice dissection of an {L'} to a square. Discovering a tessellation element for the plane, Bernard Lemaire gave an 8-piece dissection in (Berloquin, *Le Monde*, 1974a). But a Mr. Szeps sent a clever 7-piece dissection (Figure 10.29), and Lemaire then found a different 7-piece dissection; see (Berloquin,

113

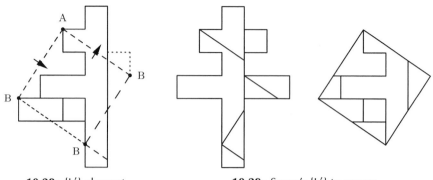

10.28: {L′} element   **10.29:** Szeps's {L′} to square

Le Monde, 1974c,d). It appears that Szeps cut the cross into three pieces to form the tessellation element shown in Figure 10.28. The outline of a square is shown with its corners anchored at the repetition points. There are three different types of edges for this square. As in the previous dissection, edges must be matched in a way that takes into account the length of dashes and the direction of arrows.

We cannot tile the plane with such an element, but we can tile the cube! Figure 10.30 shows the complete hexahedral tessellation, folded out flat. Circular arcs indicate how several of the edges fit together to form a cube, as shown in Figure 10.31. Lemaire extended Szeps's dissection to a similar dissection of the Cross of Lorraine to a Greek Cross in just seven pieces. We can also derive this dissection by using hexahedral tessellations.

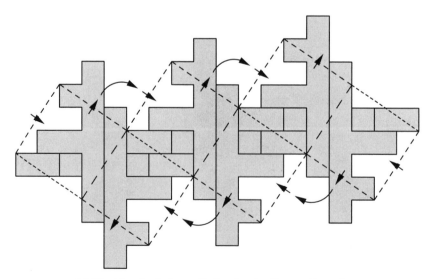

10.30: Hexahedral tessellation for the Cross of Lorraine

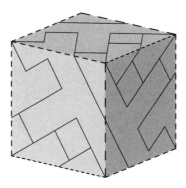

10.31: Cross of Lorraine tessellation superposed on the cube

Puzzle 10.3: Show that a Greek Cross has a hexahedral tessellation element. Find a 7-piece dissection of the Cross of Lorraine to a Greek Cross.

If these last two dissections have not taken your breath away, we will do it now, like collapsing an inflatable beach ball! We explain the dissection of two hexagrams to one from Figure 5.24 in terms of tessellating a "deflated" polyhedron: It has two hexagonal faces whose vertices and edges coincide and lie flat against each other. Thus this dihedron has a volume equal to zero. The element for the large {6/2} is given in Figure 10.32. We obtain the tessellation in Figure 10.33 by folding backward along the dotted lines. The element for the two smaller {6/2}s is given in Figure 10.34. This time we obtain the tessellation in Figure 10.35 by folding forward along the dotted lines. Superposing the two tessellations gives the dissection in Figure 5.24.

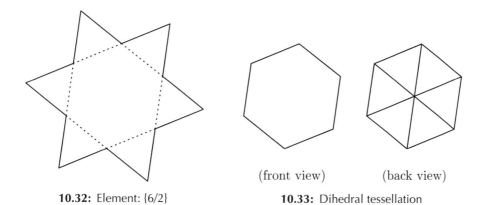

(front view)   (back view)

10.32: Element: {6/2}   **10.33:** Dihedral tessellation

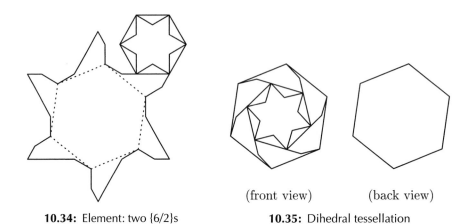

10.34: Element: two {6/2}s

(front view)   (back view)

10.35: Dihedral tessellation

We can use dihedral tessellations to derive the other dissections similar to two hexagrams to one, such as two {8/3}s to one. The dihedron in this case has two octagonal faces.

CHAPTER 11

Plain Out-Stripped

Sacrebleu! These academicians hold our lives hostage in their footnotes! First, I demonstrate to this Professeur Catalan my method by which a square can be decomposed into seven pieces that form three equal squares. He labels it "empirique" in a footnote in his book. Next I extend my method, decomposing a square into a hexagone régulier *in five pieces, and into a* pentagone régulier *in seven pieces. What does he do but suppress the constructions from my manuscript, giving as his excuse (where else but in a footnote!) "a lack of space." Finally he attaches footnote after footnote to the paper on my problem by that "Génie" M. deCoatpont. No doubt this docteur, this professeur, this membre des sociétés and associé des académies, will be the cause of my death, which he will announce, naturellement, in a footnote.*

The preceding bit of fabricated exasperation, which we assign to the Belgian Paul Busschop, underscores the lack of hoopla that accompanied his introduction of a new dissection technique in the 1870s. His first dissection was labeled as empirical by Eugène Catalan (1873), although Catalan (1879) no longer retained this characterization. In his role as journal editor, Catalan removed all but the claims of two more dissections in (Busschop 1876), due to lack of space. (However Catalan was kind enough to supply Busschop's full manuscript to Édouard Lucas, who included the dissections in (Lucas 1883).) As a matter of record, Busschop did die soon after his article appeared, and Catalan (1879) announced his death in a footnote.

The lack of recognition is unfortunate, as Busschop's technique is really nice. The basic idea is to cut each polygon into pieces that rearrange to form an element that can be repeated to fill out a planar strip of constant width and infinite length. All copies of the element in the strip must be oriented similarly. The strips for each of the figures are then "crossposed," with the lines in one strip identifying additional cuts in the figure of the other strip. Each strip is positioned so that its two boundaries cross a given boundary of the other strip at points that are at a distance equal to the length of the element of the other strip. We call this the *plain-strip*, or *P-strip*, technique.

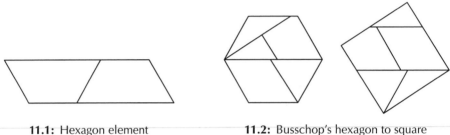

11.1: Hexagon element   **11.2:** Busschop's hexagon to square

For Busschop's dissection of a hexagon to a square, we can cut the hexagon along a diagonal into two equal trapezoids that fit together to make the parallelogram in Figure 11.1. We then line up copies of the parallelogram to form the hexagon's strip. We need make no cuts in the square to give an element, since the squares readily line up to form a strip. The crossposition of these two strips is shown in Figure 11.3. To minimize the number of pieces, Busschop had a vertex of the square coincide with a vertex of the hexagon. This is equivalent to making an intersection point in the square element coincide with an intersection point in the hexagon element. The resulting 5-piece dissection is shown in Figure 11.2.

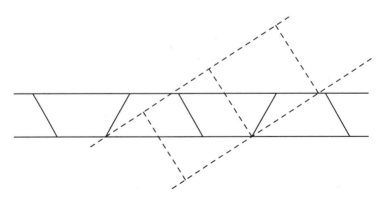

11.3: Crossposition of hexagons and squares

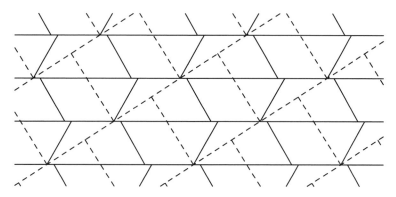

11.4: Tessellations of hexagon elements and squares

William Macaulay (1922) realized that the strip technique is a type of tessellation method. A strip gives rise to infinitely many tessellations, merely by laying copies of the strip side by side, with adjacent strips shifted relative to each other by some fixed offset. The crossposition of two strips forces a choice of offset for each strip, thus forcing a fixed tessellation for each. For the hexagon-to-square dissection, the fixed tessellations are shown superposed in Figure 11.4. The tessellation elements are the same as the strip elements.

Busschop's technique is an extension of Jean Montucla's method, in (Ozanam 1778), for dissecting a rectangle of any dimensions to a square of equal area. However, Busschop contributed two key ideas: First, he could handle elements that are not rectangular; second, he created these elements from other figures, such as hexagons, by dissection. There is no evidence that Montucla, Busschop, or others who followed them explicitly superposed strips in the manner that we illustrate. But their methods probably amounted to much the same thing.

We can also explain Busschop's dissection of a pentagon to a square in terms of P-strips. We cut the pentagon along a diagonal, giving an isosceles triangle and

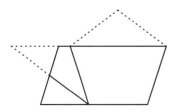
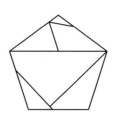
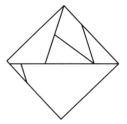

11.5: Pentagon element

11.6: Brodie's pentagon to square

a trapezoid. We then position the triangle flush against the trapezoid and cut it in two, with one piece swung around to create a parallelogram. The conversion from pentagon to parallelogram is shown in Figure 11.5, with the dotted lines indicating how to cut the isosceles triangle from the pentagon and then in two.

Busschop's pentagon-to-square dissection was not minimal, but only because he chose the wrong points from each element to coincide. Using the same elements, the Scotsman Robert Brodie (1891) gave a 6-piece dissection, which can be derived by using the same two strips. Crossposing the P-strips as in Figure 11.7 gives the dissection, shown in Figure 11.6. Dudeney (*Dispatch*, 1903a) also gave a 6-piece dissection that could be similarly derived, but Brodie seems to have found his first.

Twenty years after Brodie's dissection appeared, the Englishman William Macaulay (1915) investigated the conditions under which 8-piece dissections of one quadrilateral to another exist. We can derive Macaulay's dissection by crossposing strips for each quadrilateral (Figure 11.9). To create the element for a quadrilateral we identify a diagonal, shown with a dotted line in Figure 11.8. The diagonal divides

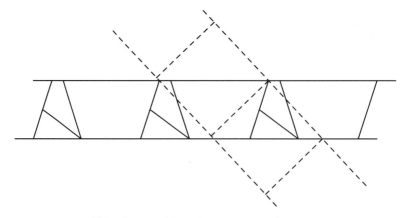

11.7: Crossposition of pentagons and squares

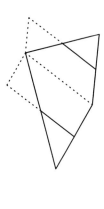

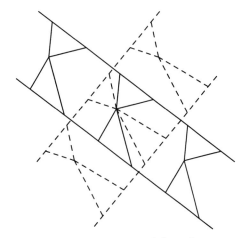

11.8: Quad element   **11.9:** Crossing quadrilaterals

the quadrilateral into two triangles. We make cuts parallel to the diagonal so that a quarter is cut off each of the two triangles. Finally we rotate the quarter-size triangles to give the element.

This is our first example of a P-strip element that is not a parallelogram. We don't need a parallelogram, since we can match up this element end-to-end to create the P-strip. The corresponding dissection is shown in Figure 11.10. Although the quadrilaterals are not symmetrical, the dissection has a hidden beauty, revealed by its cyclic hinging in Figure 11.11. Rotating the pieces as indicated by the arrows gives the quadrilateral on the right, and in the reverse direction gives the quadrilateral on the left.

From Belgium to Scotland to England, the P-strip technique was hopping around. Its next appearance was in the United States, as a clever improvement on one of Sam Loyd's dissections. Sam Loyd (*Eliz. J.*, 1908a) had given a 7-piece dissection of a hexagram to a square. His dissection resulted from cutting the top and bottom points off the star, splitting each star point in half, fitting these pieces against the

---

Robert Brodie was a civil engineer who was born in Scotland in 1861. He first attended Madras College at St. Andrews for two years, beginning in 1875. He then continued his studies as a pupil with the railroad construction firm of John Waddell & Sons, staying with them until 1895. Later joining the Nott construction firm, Brodie relocated to Westminster in 1908 and then to Bristol in 1914. Rising to be a key member of this firm, he was still at work in 1942, when he completed *The Reminiscences of a Civil Engineering Contractor*.

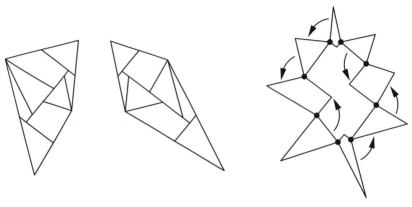

11.10: Macaulay's quad to quad   **11.11:** Cyclicly hinged

remaining portion of the star to form a rectangle, and then performing a P-slide. The appearance of Loyd's dissection was the catalyst for Harry Bradley to find the 5-piece dissection of Figure 11.13.

Bradley (1921) cut the top and bottom points off the star and nestled them in one of the two cavities of the remaining piece, giving the element shown in Figure 11.12. The crossposition of the strip of squares with this strip is shown in Figure 11.14.

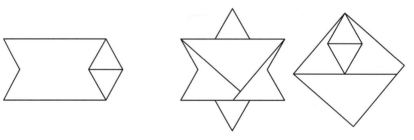

11.12: Hexagram element   **11.13:** Bradley's hexagram to square

The P-strip technique next migrated to the Southern Hemisphere, where it was embraced by the Australian Harry Lindgren. Lindgren teased and coaxed this method to solve a wide variety of dissection puzzles. Many of his dissections fell out naturally once he had identified his P-strips. We shall survey a number of these strips shortly. But first we reproduce a dissection of Lindgren's that remains a marvel no matter how many times we examine it: that of three Greek Crosses to one. Late in his career, the master Henry Dudeney (*Strand*, 1929a) had given a 13-piece dissection of three Greek Crosses to one. He had cut the three Greek Crosses into

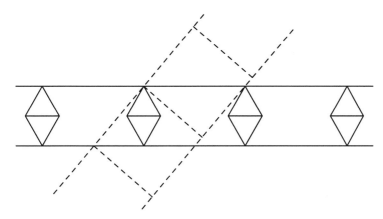

11.14: Crossposition of hexagrams and squares

pieces that form a 3×5 rectangle, which he then converted to a rectangle five times as long as it was high, using a P-slide. He then cut two squares off the rectangle and positioned them above and below the remaining rectangle to give the large Greek Cross.

Puzzle 11.1: Construct a 13-piece dissection of three Greek Crosses to one that uses a P-slide.

Remarkably, Lindgren (1961) was able to reduce the number of pieces by one. Lindgren crossposed strips for one Greek Cross and for three Greek Crosses, as shown in Figure 11.16. Although the choice for the latter strip is natural, the choice

William Herrick Macaulay was born in Hodnet, Salop, England, in 1853. He received a B.A. from Durham University in 1874 and an M.A. in mathematics from King's College, Cambridge, in 1877. Elected to a fellowship at King's, he worked in a granite quarry company until his appointment as college lecturer in 1884. Macaulay assisted in the development of the School of Engineering, was appointed second bursar in 1887, and served as tutor from 1902 until his retirement in 1913. He returned from retirement to serve as vice provost from 1918 to 1926.

Macaulay wrote two texts for engineering students, on thermodynamics and on solid geometry. He was the last fellow at King's to keep a horse and rode to the hounds up until the First World War. He was described from his quarry days as "a very nice gentleman, but he was a terrible man to tell half a story to." He died at Clent in 1936.

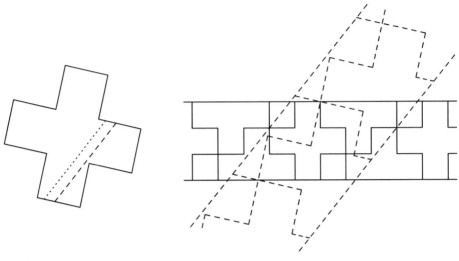

11.15: Sample cuts

11.16: Cross crosses crosses

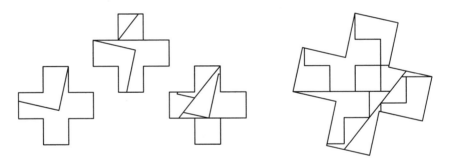

11.17: Lindgren's three Greek Crosses to one

for the former takes more care. There is an infinite family of similar strips, each found by cutting diagonally through the cross. Two sample cuts are shown in Figure 11.15. The dashed cut leads to the strip in Figure 11.16. This strip allows the two intersections of line segments in one strip to fall on top of lines and intersections in the other strip. The dotted cut leads to a strip for which there is no crosspositioning that reduces the number of pieces to 12. The dissection itself is in Figure 11.17.

We call this technique of choosing the best strip from an infinite family of strips the *optimizing strip* technique.

But, surprisingly, pushing Dudeney's approach a step further also gives a 12-piece dissection:

Puzzle 11.1, revisited! If you found a 13-piece dissection of three Greek Crosses to one, see whether you can extend it to a 12-piece dissection.

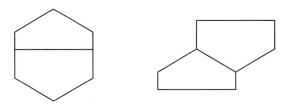

11.18: One of (infinitely) many hexagon elements

Harry Lindgren made an extensive survey of P-strips for various figures. Some of his P-strip elements appear in the following figures. He identified an infinite family of elements for the hexagon, which result from cutting a line perpendicularly through two opposite sides. One example is shown in Figure 11.18, with the cut on the left and the resulting P-strip element on the right. Lindgren (1964b) illustrated this approach with two examples: The first positioned the perpendicular cut as high as possible, and the second positioned it in the middle. The former yields an economical dissection of a hexagon to a pentagon.

Puzzle 11.2: Find a crossposition for an economical dissection of hexagon and pentagon.

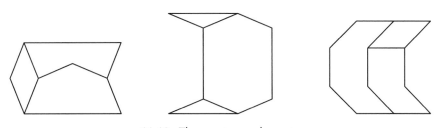

11.19: Three octagon elements

The octagon has several possibilities for P-strips, again as demonstrated by Lindgren (1951) and (1964b). Three of the elements are illustrated in Figure 11.19. (We have not shown how to cut the pieces from the octagon in each case: Try reassembling the octagons.) A priori, it is usually not clear which element we should

125

use to create the most economical dissection. The middle element in Figure 11.19 may often produce the best results, since it has a large area in one piece and relatively small pieces otherwise. A good strategy when crossposing two strips is to try to fit all the small pieces of one element inside a large piece of the other, and vice versa.

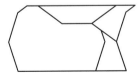

11.20: Decagon element

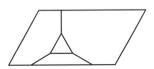

11.21: Dodecagon element

Lindgren (1964b) discovered the elements for the decagon and dodecagon that are shown in Figures 11.20 and 11.21. Using these, the reader can find a variety of interesting dissections.

Puzzle 11.3: Find a crossposition for an economical dissection of hexagram and dodecagon.

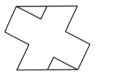

11.22: Greek Cross elements

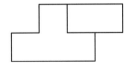

11.23: {L} element

We have already seen one P-strip element for the Greek Cross. Two more elements for it, from Lindgren (1951) and (1964b), are given in Figure 11.22. A simple element for the Latin Cross, from Lindgren (1951), is given in Figure 11.23.

Puzzle 11.4: Find a crossposition for an economical dissection of a Latin Cross to an octagon.

One dissection that turned out to be surprisingly challenging was a dissection of a pentagram to a square. Lindgren (1958) gave an 8-piece dissection, by cutting the star into a 36°-rhombus and three other pieces that could be assembled into a parallelogram. He converted the rhombus and parallelogram to rectangles of the same width as the desired square, thereby producing two more pieces each. He

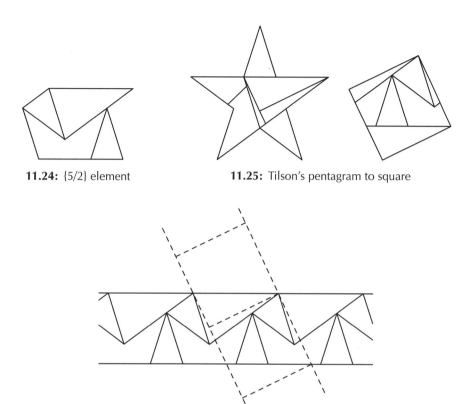

11.24: {5/2} element

11.25: Tilson's pentagram to square

11.26: Crossposition of pentagrams and squares

then stacked the two rectangles to give a square. We do not illustrate Lindgren's dissection because in this case the master of P-strips was beaten!

Philip Graham Tilson, from Wales, found a way to cut the pentagram into four pieces that form a P-strip element, as shown in Figure 11.24. This strip works well when the strip for squares is crosspositioned over it, as shown in Figure 11.26. The resulting 7-piece dissection from (Tilson 1979) is shown in Figure 11.25.

Henry Dudeney (*Strand*, 1927b) posed the puzzle of dissecting a heptagon to a square and gave a 10-piece solution by George Wotherspoon. He also gave a further challenge to his readers, of trying to solve the problem in nine pieces. Subsequently, Dudeney (*Strand*, 1927c) wrote,

> My invitation to readers to attempt a dissection in nine pieces has resulted in my receiving correct and ingenious solutions from Lt.-Col. C. E. P. Sankey and from Mr. H. G. Charlton. I regret that the limitations of space prevent my giving their methods. They are difficult and a little complex, but I think the

127

solution of the first-named gentleman is the better of the two. Their record will be hard, perhaps impossible, to beat.

Because neither of these dissections is included in any of Dudeney's books, they are probably lost. Lindgren (1964b) found three different 9-piece dissections of a heptagon to a square. But all of these efforts have been out-stripped not once, but twice!

First Anton Hanegraaf, of the Netherlands, partitioned the heptagon into four pieces that formed a tessellation element. Taking one more piece, he created a plain-strip element, which gave an 8-piece dissection of a heptagon to a square. My response to his wonderful dissection was "Now, *this* record will be hard, perhaps impossible, to beat!" But two months before the final manuscript of this book was due, Gavin Theobald, of England, sent me a remarkable set of dissections. It included a totally unexpected dissection in seven pieces!

Theobald based his dissection on several neat ideas. First, he partitioned the heptagon to form a plain-strip element that has only four pieces, as shown in Figure 11.27. The small triangular piece not only fills out the boundary of the element but leaves a cavity into which he rotates the long, thin piece. The strip gives improvements for many dissections, such as to a pentagon in nine pieces, to a hexagon in eight pieces, to an octagon in eleven pieces, to a decagon in eleven pieces, to a {G} in nine pieces, and to a {L} in eight pieces. These improved those of Hanegraaf, which had improved those of Lindgren! But the element doesn't immediately give an improvement for the heptagon to the square.

Theobald's second idea was to find a line segment perpendicular to one of the sides and of length equal to the side length of the desired square. This line segment is shown as a dashed line in Figure 11.28, with the portion of the element to its

George Wotherspoon was born in 1854 at Camberwell, Surrey, England. He studied at Trinity College in Oxford, earning a B.A. in 1876 and an M.A. in 1879. Appointed as an assistant master of King's College School in London in 1878, he was eventually responsible for much of the teaching in classics and mathematics that prepared the pupils for scholarships at Cambridge and Oxford. Before retiring in 1913, he had become vice master.

Wotherspoon enjoyed a second career of sorts, embracing a continuing interest in mathematics. He was one of the "band of experts" who regularly solved Dudeney's puzzles in *The Weekly Dispatch*. Between 1928 and 1941 he published four Mathematical Notes, primarily on geometry, in *The Mathematical Gazette*. In 1935, as an octogenarian, he joined the Mathematical Association of Great Britain.

right shifted over to the left, as shown by the dotted lines. The result is shown in Figure 11.29 in solid lines, expanded to form a parhexagon by adding the area bounded by the dotted lines. We can perform an H-slide on this expanded hexagon, with the additional cuts shown with dashed lines. Note that a part of one of the H-slide cuts coincides with one of the existing line segments. When we discard the added area, a square remains. A tour de force of dissection techniques, the complete 7-piece dissection is shown in Figure 11.30. Can we finally assert?: Surely, *this* record will be impossible to beat!

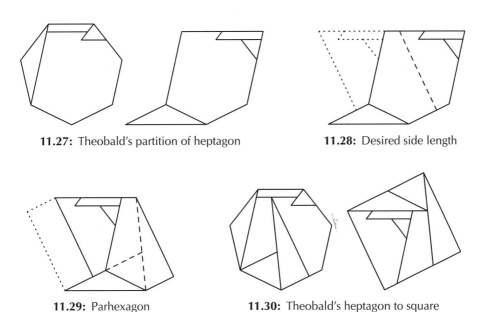

11.27: Theobald's partition of heptagon   **11.28:** Desired side length

11.29: Parhexagon   **11.30:** Theobald's heptagon to square

Besides heptagons, both Hanegraaf and Theobald set their sights on enneagons and handily outstripped Harry Lindgren in the process. Lindgren (1964b) gave 12-piece and 14-piece dissections of an enneagon to a square and to a hexagon, respectively. But by designing a clever P-strip element, Hanegraaf reduced these by two and three pieces, respectively. David Paterson (1995b) also gave a 10-piece dissection of an enneagon to a square, and a 12-piece dissection to a hexagon. Hanegraaf's P-strip element is produced by using a new technique, which is similar to Lindgren's Q-slide technique. It converts a trapezoid to a parallelogram, so that I call it a *trapezoid slide*, or T-slide.

An example of a T-slide is shown in Figure 11.31. Starting with the trapezoid ABCD on the left, we locate point E on AB so that the length of AE is the length of the desired parallelogram. Next we locate point G on CD so that the sum of the lengths of BE and GC equals the length of AE. Finally, we let point F be the midpoint of DG and H the midpoint of BC. We cut along FH, then cut down from E parallel to AD and up from G parallel to AD. The T-slide can be cyclically hinged, as shown in Figure 11.31.

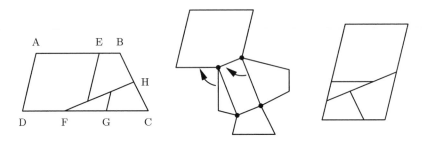

11.31: Hanegraaf's T-slide

To create his element, Hanegraaf cut an enneagon as on the left of Figure 11.32 and rearranged the pieces as indicated by the solid lines in the middle. The portion of the middle figure below and/or to the right of the dotted line is a trapezoid to which a T-slide can be applied. The dashed lines indicate the cuts of the T-slide,

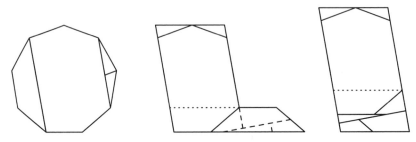

11.32: Hanegraaf's enneagon element

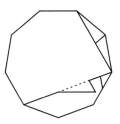 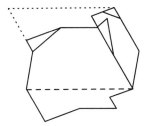

11.33: Partition {9}   **11.34:** Theobald's element

and the parallelogram on the right of Figure 11.32 shows the result of the T-slide. To form a P-strip, just rotate the element 90° and fit like copies end-to-end. Besides dissections to square and hexagon, Hanegraaf identified a number of others by using this strip.

Although Hanegraaf's enneagon element is nifty, Gavin Theobald produced a strip element that can give better results. He first partitioned the enneagon into five pieces, as shown in Figure 11.33. The dotted line indicates a triangular out-cropping from the largest piece into the long piece beneath. When we rearrange the pieces into a tessellation element in Figure 11.34, the inclusion of this out-cropping guarantees that the large piece is adjacent in the tessellation to another copy of itself. In the rearrangement, the vertices at the lower right of the largest piece and the long piece do not exactly coincide. This is because the length of the long piece, $1 + 2\cos 40° - 1/(4\cos 40°) \approx 2.206$ (where the side of the ennea-gon is assumed to be 1), is just slightly less than the length of the two edges, $2\sin 10° + 2\cos 20° \approx 2.227$, against which it is placed.

Theobald converted his tessellation element to a strip element by cutting along the dashed line, which spans from the rightmost vertex of the largest piece to

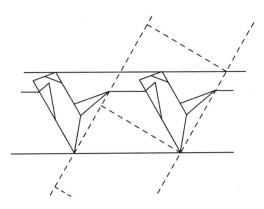

11.35: Theobald's {9} and {4} crossposition

approximately 2.227–2.206 below the leftmost vertex of the largest piece. The dotted lines show how to reposition the piece beneath the dashed line to give the strip element.

The crossposition of enneagon and square is shown in Figure 11.35, and the resulting 9-piece dissection is shown in Figure 11.36. The circles indicate pairs of distinct vertices that are so close together as to appear coincident in the figure. The enneagon strip gives an 11-piece dissection to a hexagon, the same number of pieces as Hanegraaf's.

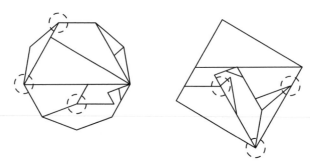

11.36: Theobald's enneagon to square

Whereas Theobald's heptagon and enneagon strips are general-purpose strips, he designed other strips for specific dissections, such as an octagon to a hexagon. Lindgren (1964b) gave a 9-piece dissection, but Theobald has gone one better by using the optimized strip approach.

Theobald started with Lindgren's (1964b) tessellation element for an octagon (Figure 11.37). There are an infinite number of possible cuts that will transform it into a strip element, with two examples shown with the dashed line and with the

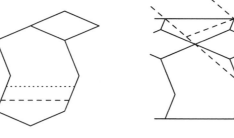

11.37: Sample cuts

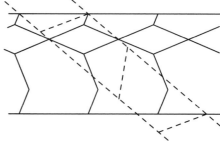

11.38: Crossposing octagons and hexagons

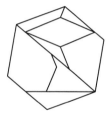

11.39: Theobald's octagon to hexagon

dotted line. Theobald crossposed a hexagon strip based on Figure 11.1 to avoid cutting the octagon's rhombus. To have the boundaries of the hexagon strip cross the octagon strip at intersection points of line segments, he used the dashed cut in Figure 11.37 to produce the octagon strip. The crossposition of strips is shown in Figure 11.38, and the resulting 8-piece dissection is shown in Figure 11.39.

An exciting feature of Gavin Theobald's work is his invention of a technique that I call customized strips. A *customized strip* is derived from a crossposition for a specific dissection, with the strip altered so that certain of its line segments coincide with line segments in the other strip. We illustrate with Theobald's marvelous 7-piece dissection of a decagon to a square.

Theobald first partitioned a decagon into four pieces and found the general-purpose decagon strip element, as shown with solid and dotted lines in Figures 11.40 and 11.41. Then he crossposed this strip with a strip of squares and discovered that the small quadrilateral on the right that is bounded by dashed and dotted lines could be attached to the lower right piece in the decagon partition. Of course, he had to cut a hole into which this attachment would fit in the strip element, but he attached the piece from this hole to the isosceles triangle, which then filled out a hole in the strip element.

The new partition and customized strip element are shown with dashed and solid lines in Figures 11.40 and 11.41. The crossposition of square strip and customized decagon strip is shown in Figure 11.42, and the resulting dissection, which

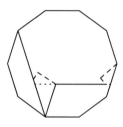

11.40: Partition of {10}

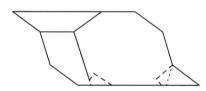

11.41: Decagon element

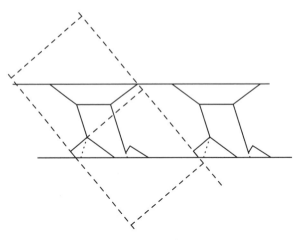

11.42: Theobald's decagon and square crossposition

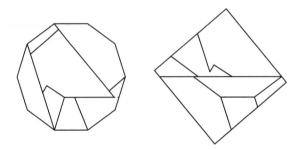

11.43: Theobald's decagon to square

improves on an 8-piece dissection by Lindgren (1964b), is shown in Figure 11.43. How many different strip elements did Theobald examine to find one that permitted the advantageous swap of quadrilaterals?

A remarkable adaptation of the plain-strip technique was discovered by Robert Reid, of Lima, Peru. He (1987) found a 9-piece dissection of a {12/2} to a square, using what we identify as a *bumpy plain strip*. A plain strip for {12/2}, in a combination of dotted and solid lines, is shown in Figure 11.44, along with a crossposition of a strip of squares, in dashed lines. The bumpy strip is outlined in only solid lines. Using the nonbumpy strip indicated by solid and dotted lines gives a 10-piece dissection.

But the bumpiness is the creative part. In the tessellation induced by the crossposition, the outward bumps on one strip precisely match the inward bumps on another. In fact, Reid cut the {12/2} into pieces to position the bumps so that they would match up when the dashed lines were laid down. Flexibility in the placement

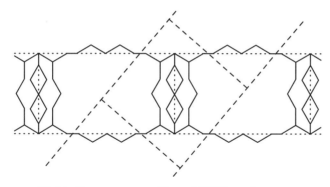

11.44: Bumpy crossposition of {12/2}s and squares

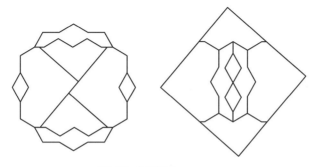

11.45: {12/2} to square

of the bumps allows the technique to be successful in this case. Since the outward bumps are not centered in the large piece, the piece cut off the top of the {12/2} must be turned over before it can be matched up with the piece from the bottom of the {12/2}. The crossposition shown in Figure 11.44 is a variation of Reid's found by Gavin Theobald, which reduces the number of pieces to eight, as shown in Figure 11.45. Evidently, neither Robert Reid nor I learned the lesson of appropriate crossposing that Robert Brodie supplied to Paul Busschop. Is this perhaps a case of "Sacrebleu all over again"?

CHAPTER 12

Strips Teased

"I exhibited this problem before the Royal Society, at Burlington House, on 17th May 1905, and also at the Royal Institution in the following month, in the more general form: – 'A New Problem on Superposition. . .' I add an illustration showing the puzzle in a rather curious practical form, as it was made in polished mahogany with brass hinges for use by certain audiences. . ." Such was Henry Ernest Dudeney's legendary introduction of the magnificent 4-piece dissection of a triangle to a square.

What excitement! And yet lurking beneath the surface is a sense of being teased by what we are not told: How was Dudeney received in those rarefied (and perhaps stuffy) circles? What became of his paper "A New Problem. . ."? Who were those certain audiences? Where did that polished mahogany model end up? How was the dissection discovered?

The triangle-to-square dissection, shown in Figure 12.2, and in fully hinged form in Figure 12.1, is most remarkable. The puzzle was originally posed by Dudeney (*Dispatch*, 1902a), followed two weeks later with some unusual discussion by Dudeney (*Dispatch*, 1902b), and finally two weeks after that with the solution and explanation by Dudeney (*Dispatch*, 1902c). A careful study of the three columns can produce some doubt as to whether Dudeney was the one who discovered the dissection. We are teased by the delay of an additional two weeks in publishing the solution; by the generous praise that Dudeney heaped on C. W. McElroy, the only reader to solve the problem in four pieces; and by the failure of Dudeney to state unequivocally that he had discovered the 4-piece dissection. Could it have been discovered by McElroy?

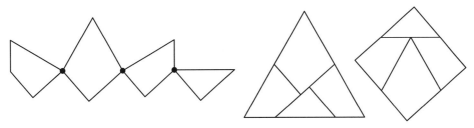

12.1: Hinged pieces   **12.2:** Triangle to square

Subsequently, Dudeney (1907) discussed the solution in the form from which the preceding excerpt was taken. We can also feel teased that there seems to be no independent record of the "royal" events, neither in the *Proceedings of the Royal Society* nor in the *Proceedings of the London Mathematical Society.*

The dissection was probably not discovered by the technique of this chapter, although it is the earliest example that could have been. This technique is an ingenious variation of the P-strip technique of the last chapter. As before, we cut a figure into pieces that form a repeating element within a strip. But now we rotate every second element 180° and then match it with an unrotated "twin." Our strip is then a sequence of pairs of twinned elements. Every two consecutive elements in such a strip share a point of 180°-rotational symmetry at the middle of their boundary. We call such a point an *anchor point.* When we crosspose two strips, we must either cross any anchor point in the intersection of the two strips by a boundary of the other strip or cover it by an anchor point in the other strip. We call this method the *twinned-strip,* or *T-strip,* technique.

In the triangle-to-square dissection, the T-strip for the triangle merely takes the triangle as the element. The midpoint of every side shared by two triangles is an anchor point. The strip for squares is also easily formed, and the two strips are crossposed in Figure 12.3. To achieve the crossposition, we must satisfy a curious condition. The overlap of the two strips must cover exactly the area of two triangles (and equivalently of two squares). This forces the strip of squares to be treated as a second T-strip. Tiny circles indicate the anchor points of both strips.

Clearly, the use of T-strips is more restricted than the use of P-strips. Depending on the relative shape of the T-strip elements, several variations are possible. As in the preceding example, two T-strips may overlap on double the area of the element. Harry Lindgren (1964b) called this a *TT2* dissection. When the area of overlap of the T-strips exactly equals the area of the element, Lindgren called it a *TT1* dissection. When a P-strip and a T-strip are used, Lindgren called it a *PT* dissection. In compensation for the restrictiveness of T-strips, anchor points usually give rise to locations where pieces can be hinged. This seems to happen whenever

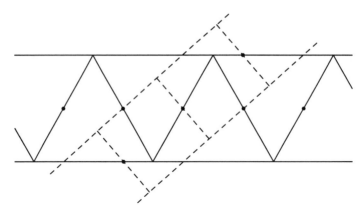

12.3: Crossposition of triangles and squares

a piece in one strip is cut by an edge in the other strip and there is an anchor point somewhere along this cut.

We will be teased by many features of these dissections. For example, we can fully hinge the triangle-to-square dissection in more than one way. If we remove any one of the three hinges in the chain shown in Figure 12.1, then we can connect with a hinge the right vertex of the triangle (on the right) with the middle left vertex of the quadrilateral on the left. For example, if we replace the rightmost hinge in Figure 12.1, we get the chain shown in Figure 12.4, where the triangle has been shifted from the dotted position on the right to its new position on the left. Surprisingly, no one seems to have noticed this before. Did everyone miss the fact that the legendary model used by Dudeney before certain audiences could have been hinged in three other ways?

Henry Taylor (1905) pointed out that the dissection of a square and an equilateral triangle is a special case of a more general dissection of a nonequilateral triangle and a parallelogram. Did Taylor make his discovery independently? At the time that his paper was published, Taylor was blind, so he may have made the

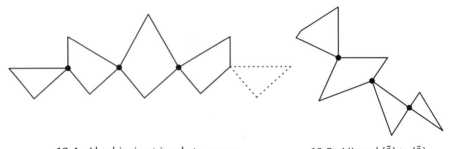

12.4: Also hinging triangle to square   **12.5:** Hinged {3̄} to {3̄}

discoveries for his paper much earlier. Actually, a more general version of the dissection is possible, namely, of a nonequilateral triangle and a trapezoid. This has been used already in Figure 5.9. Of course, the general dissection does not apply to all combinations of shapes of nonequilateral triangles and trapezoids.

Taylor recognized that a similar approach could be applied to the problem of dissecting a nonequilateral triangle to a different nonequilateral triangle, if their dimensions were not too different. Lindgren (1953) identified the crossposition of strips in Figure 12.6 that corresponded to Taylor's dissection in Figure 12.7. There is no evidence that Taylor crossposed strips, however. This dissection is another example of a TT2 dissection, and we can fully and variously hinge it. A sample hinging is shown in Figure 12.5. If we swing the leftmost piece clockwise as far as it will go, we get the top triangle in Figure 12.7. If we swing it counterclockwise, we get the bottom triangle. Taylor also gave an example in which the triangles were sufficiently dissimilar that the method fails to give a 4-piece dissection.

We have seen an 8-piece dissection of two quadrilaterals in the last chapter, and now a 4-piece dissection of two nonequilateral triangles. What about a quadrilateral to a nonequilateral triangle? Lindgren (1964b) found the crossposition of strips shown in Figure 12.8 and the corresponding 6-piece TT2 dissection in Figure 12.9. The quadrilateral T-strip element is similar in spirit to the P-strip element in Figure 11.8. But since two are placed back to back, one of the two triangles need not be cut off and flipped around as in Figure 11.8.

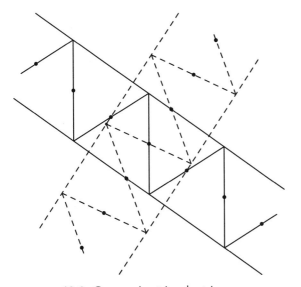

12.6: Crossposing triangle strips   **12.7:** Taylor's triangles

Henry Martyn Taylor was born at Bristol, England, in 1842. He studied mathematics at Trinity College in Cambridge, earning a B.A. degree in 1865 and an M.A. in 1868. After four years as a vice principal of the Royal School of Naval Architecture, and despite having been called to the bar at Lincoln's Inn, he returned to Trinity in 1869 as assistant tutor on the mathematics staff. He later served as tutor and retired in 1894. Taylor's main research interest was in geometry.

Soon after retirement, Taylor suffered a severe attack of influenza that rendered him totally blind by 1897. Mastering the Braille system and recognizing that there was little scientific material available in Braille, Taylor transcribed a series of scientific textbooks. He focused on extending Braille to accommodate mathematical notation and diagrams. In this later period of his life, he also served as an alderman for Cambridge and as mayor. He died at Cambridge in 1927.

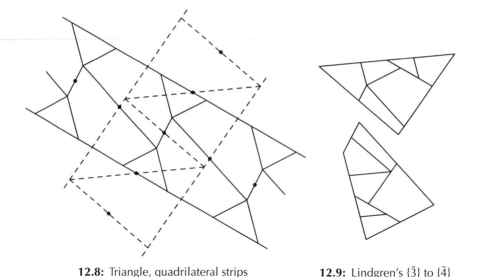

12.8: Triangle, quadrilateral strips

12.9: Lindgren's {3̄} to {4̄}

With two different ungainly polygons, it is at first difficult to get excited about this dissection. But a closer look reveals a hidden beauty: There is a nifty way to hinge it.

Puzzle 12.1: Give as complete a hinging as possible for Lindgren's quadrilateral to nonequilateral triangle dissection.

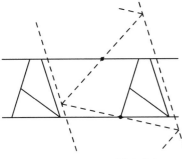

12.10: Crossposed {5}, {3}

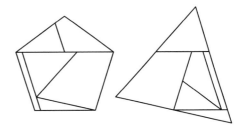

12.11: Goldberg's pentagon to triangle

Although we have dissected triangles to squares, to quadrilaterals, and to other triangles, we discover that we have not yet satisfied the "pent-up" demand for triangle dissections. Michael Goldberg (1952) gave a 6-piece dissection of a pentagon to a triangle, shown in Figure 12.11. Lindgren (1964b) showed how to derive this dissection by crossposing the P-strip element for pentagons in Figure 11.5 and the T-strip element for triangles, as shown in Figure 12.10. We can partially hinge it, forming two chains rather than one. One chain connects the top two pieces of the dissected pentagon in Figure 12.11. We can variously hinge the remaining four pieces. This PT dissection matches a 6-piece dissection by A. H. Wheeler mentioned in (Ball 1939) but apparently never published.

Let's put the "hex" on the triangle. Michael Goldberg (1940) gave a 6-piece dissection of a hexagon to a triangle as a problem in the *American Mathematical*

Michael Goldberg was born in New York City in 1902, the first American-born child of Polish immigrants. He was raised in Philadelphia and earned a B.S. in electrical engineering from the University of Pennsylvania in 1925. After a year with the Philadelphia Electric Company, he joined the Navy Department's Bureau of Ordnance in Washington, D.C. Evening studies led to an M.A. in mathematics from George Washington University in 1929. For the navy he worked on gunfire control systems, and eventually missile fire control systems, until he retired in 1963 from what had become the Bureau of Naval Weapons.

Mathematics was Goldberg's principal hobby. He published over 60 papers in areas such as dissection problems, mechanical linkages, plane and solid rotors, and packing problems. Each summer, he planned the family vacation as a trip to meetings of the Mathematical Association of America and other organizations. Michael Goldberg died in 1990.

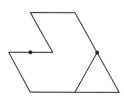

12.12: Hexagon element

12.13: Lindgren's hexagon to triangle

Monthly. When he received two other 6-piece dissections, the problem editor challenged readers to find a 5-piece dissection. It took a decade, but Lindgren (1951) found such a 5-piece dissection (Figure 12.13). Creating the T-strip element of Figure 12.12, he crossposed two T-strips as in Figure 12.14. We can partially hinge the dissection, with all but the small triangular piece in one chain, and also variously hinge it.

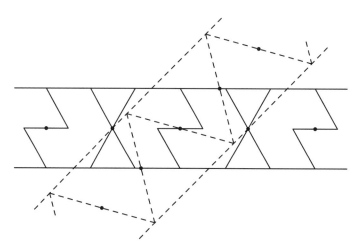

12.14: Crossposition of hexagons and triangles

If you feel "crossed-up" with the last several triangle dissections, you must be in the mood for the Greek Cross. Half a year after the triangle-to-square dissection, Dudeney (*Dispatch*, 1902e) found out how to modify it to give a 6-piece dissection of the Greek Cross to a triangle. Although he probably didn't use T-strips, we can use them to produce his dissection. Find the one Greek Cross element in the infinite family suggested by Figure 11.15 that forms a T-strip. Then superpose this with the T-strip for the triangle to give a TT2 dissection. A nice feature of this dissection is that we can cyclicly hinge four of the pieces and connect the other two to this cycle by hinges.

Puzzle 12.2: Find the one T-strip element in the infinite family of Greek Cross elements suggested by Figure 11.15. Form the TT2 dissection with the triangle and hinge the dissection.

Surprisingly, Dudeney was beaten once again. Harry Lindgren (1961) used his (1951) T-strip element for the Greek Cross to give a different TT2 dissection, producing the crossposition of the two T-strips in Figure 12.15 and the 5-piece dissection shown in Figure 12.16. Although Lindgren saved one piece over Dudeney's dissection, his dissection cannot be cyclicly or even fully hinged. But we can variously hinge all pieces except the small square.

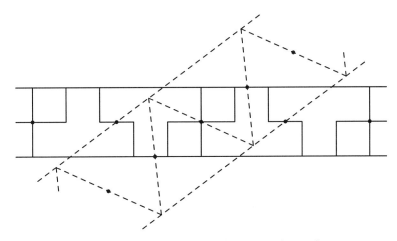

12.15: Crossposition of Greek Cross and triangle

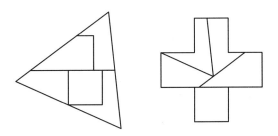

12.16: Lindgren's Greek Cross and triangle

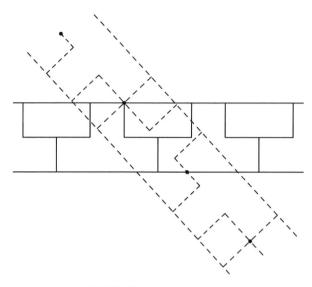

12.17: Cross crosses cross

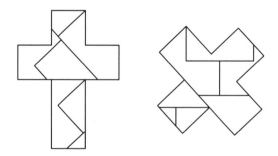

12.18: Lindgren's {L} to {G}

Having picked up the Greek Cross, Lindgren (1953) could not resist "crossing the Rubicon" and simultaneously snatching the Latin Cross. He dissected one cross to the other, finding that a T-strip works better for the Greek Cross. His crossposition of strips (Figure 12.17) leads to his 7-piece PT dissection (Figure 12.18).

Lindgren (1964b) reported that Ernest Irving Freese had found a 9-piece dissection of an enneagon to a triangle. Frustrated after considering a variety of alternatives, Lindgren concluded that beating Freese would require finding a needle in a haystack. Robert Reid found that needle, with an elegant dissection that uses the Q-slide. However, he needed to turn over one of the pieces. So back at the haystack, Gavin Theobald found a different 8-piece dissection, and with no pieces turned over.

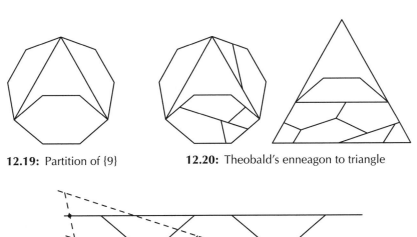

12.19: Partition of {9}   **12.20:** Theobald's enneagon to triangle

12.21: Crossposing for trapezoids

Albert Harry Wheeler was born in Leominster, Massachusetts, in 1873. He received a bachelor's degree from Worcester Polytechnic Institute in 1894 and a master's degree from Clark University in 1921. A charter member of the Mathematical Association of America, he taught mathematics at North High School in Worcester, Massachusetts. Wheeler wrote several textbooks on algebra and also lectured on mathematics. At the International Mathematical Congress in Toronto in 1924, he presented 22 of the 59 stellations of the icosahedron, of which 10 were not previously known. Harry Wheeler died in 1950.

Theobald started with Freese's basic idea, of cutting three trapezoids off the enneagon to leave a revealed triangle. Freese used a T-strip dissection to convert the three trapezoids to a larger trapezoid that forms the base upon which the revealed triangle rests. To reduce the number of pieces, Theobald nibbled a fourth trapezoid away from the revealed triangle and merged the new trapezoid with the one beneath it, to give the partition in Figure 12.19. The nibbled triangle plus a trapezoid gives a triangle that rests upon a larger trapezoid. Then Theobald could create the larger trapezoid from a T-strip element that contains just two pieces. The crossposition is shown in Figure 12.21, and the resulting 8-piece dissection is in Figure 12.20. Haystacks can hold most elegant needles!

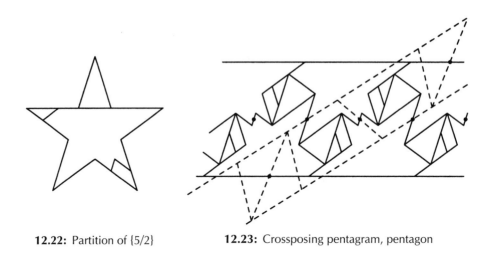

12.22: Partition of {5/2}

12.23: Crossposing pentagram, pentagon

12.24: Theobald's pentagram to pentagon

Lindgren (1964b) produced a large number of dissections using T-strips, inevitably prompting the question, Could any of them be beaten? In answer, Gavin Theobald teased the most remarkable strips from polygons and stars and produced some stunning dissections. So many, in fact, that without a doubt he is now the new master of strips. We start with a pentagram to a pentagon, for which Lindgren (1964b) gave an approximate dissection. Tilson (1979) gave an exact 10-piece dissection that used his pentagram strip from Figure 11.24.

But Gavin Theobald produced a marvelous 9-piece dissection, based on an inventive T-strip element. First he partitioned the pentagram as in Figure 12.22. He crossposed this corresponding strip in Figure 12.23 with a pentagon T-strip from (Lindgren 1964b). The resulting dissection is given in Figure 12.24. Theobald also used his pentagram strip to improve on dissections to a triangle and to a hexagon.

Theobald produced an impressive T-strip element for the octagon, whose partition and rearrangement are shown in Figure 12.25. Fitting a small triangle atop one

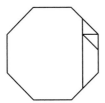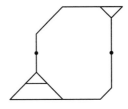

12.25: Octagon partition and T-strip element

of the pieces is reminiscent of one of the tricks he used in Figure 11.27. It seems likely that no other 4-piece partition leaves as much area in one large piece. The strip makes possible a 7-piece dissection to a triangle, beating Lindgren (1964b) by one piece. Combined with Theobald's heptagon strip, the resulting 11-piece dissection beats Lindgren (1964b) by two pieces and Hanegraaf by one.

Puzzle 12.3: Use Theobald's octagon strip to find a 7-piece dissection into a triangle.

In the last chapter, we first saw Gavin Theobald's technique of customizing strips. Now he will tease us almost into disbelief with his dissections of {8/3}. Lindgren (1964b) produced a tessellation element for an {8/3} by slicing off one piece containing both points on the top, and another piece containing both points on the bottom. Splitting both of these pieces in half gives a P-strip element, but Theobald realized that splitting just one of the pieces, say the piece from the top, gives rise

Gavin Alexander Theobald was born in London, England, in 1961. He graduated with a B.Sc. with honors in computer studies at Loughborough University of Technology in 1983. Since 1984 he has worked as a computer programmer for Computer Concepts Ltd. and its successor, Xara Ltd. Most recently he has specialized in low-level high-performance graphics rendering engines that are at the heart of some award-winning software.

Gavin enjoys card and board games and has many mathematical interests for which he uses the computer as a tool. He has been interested in regular polyhedra since he was a child and was drawn into geometric dissections in 1988 after reading one of Martin Gardner's books. He and his wife enjoy mountain climbing and camping trips to remote mountain glens in Scotland. They are also active in church activities.

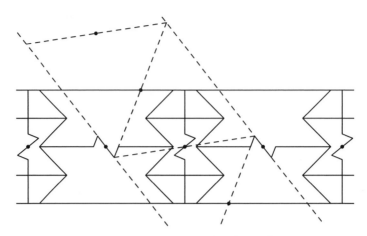

12.26: Crossposing {8/3} and triangle

to a T-strip element. Furthermore, the piece at the bottom can be severed with a cut that has 2-fold rotational symmetry about its midpoint.

Thus Theobald first took the simple T-strip element, with the cut being a straight line segment, and crossposed it with a strip of triangles. Then he altered the cut on the {8/3} to conform with some of the edges in the triangle strip. The resulting {8/3} strip and crossposition are shown in Figure 12.26. The 8-piece dissection is shown in Figure 12.27.

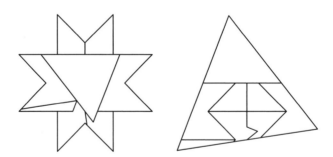

12.27: Theobald's {8/3} to triangle

In the same freewheeling spirit, Theobald created an amazing dissection of an {8/3} to a {6/2}, which I cannot resist including. It also uses a customized strip for the {8/3}, and in addition uses a different, fatter strip for {6/2} than the one introduced in the last chapter. The crossposition is shown in Figure 12.28, and the surprising 10-piece dissection is given in Figure 12.29.

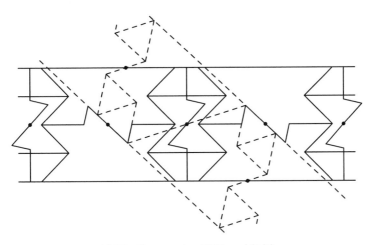

12.28: Crossposing {8/3} and {6/2}

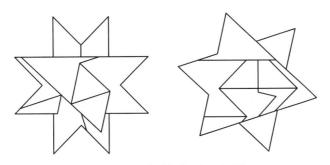

12.29: Theobald's {8/3} to {6/2}

Gavin Theobald's final teaser is a 7-piece dissection of a decagon to a triangle. It uses a customized strip based on the general-purpose decagon strip suggested in Figure 11.41.

Puzzle 12.4: Find a 7-piece dissection of a decagon and triangle, based on a customized T-strip for a decagon.

CHAPTER 13

Tessellations Completed

For many, it was a highlight of the month when The Strand Magazine *appeared. Besides articles of general interest and the short stories by such notables as G. K. Chesterton, P. G. Wodehouse, and A. Conan Doyle, there was "Perplexities," the all-too-aptly-titled column of mathematical puzzles by that clever chap Henry E. Dudeney. In the May 1926 issue, Dudeney had solved a difficult puzzle, that of cutting a regular octagon into a square in only seven pieces. Then, in September, he concluded his column with the following unexpected comment:*

"As I have often remarked, concerning these Cutting-out puzzles, there is generally no way of determining positively that you have discovered the most economic solution possible. The solution to this puzzle, in seven pieces, that I gave in our May issue, was the best then known, but the record has been beaten, for I have received from Dr. G. T. Bennett this very simple and elegant solution in as few as five pieces. I do not think that this can possibly be beaten."

Geoffrey Bennett's dissection, from (Dudeney, *Strand*, 1926d), is shown in Figure 13.2. It is one of the most beautiful in the field. As the excerpt establishes, the dissection is also a notable example in which the master puzzlist Henry Dudeney was bettered. Dudeney's 7-piece dissection, in (Dudeney, *Strand*, 1926b), transformed the octagon into a rectangle in four pieces and then converted the rectangle to a square with a P-slide, introducing three more pieces. But Bennett's was a new type of dissection, using a novel technique that Harry Lindgren has called *completing the tessellation.*

As we have seen in Chapter 10, polygons that are tessellation elements or that we can transform easily into tessellation elements often have economical dissections to other polygons. But what happens when a polygon is not a tessellation element or cannot be transformed easily into a tessellation element with a desired repetition pattern? It is sometimes possible to create a tessellation element by the addition of another polygon. If this additional polygon can also be added to the target polygon to form a tessellation element with the same repetition pattern, then an economical dissection may result from superposing the two tessellations.

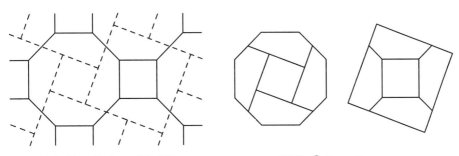

13.1: Tessellations: {8}, {4}   **13.2:** Octagon to square

As shown in Figure 2.12, an octagon and a square of equal side length form a tessellation element with a repetition pattern that is a square. Furthermore, any two squares form a tessellation element with a repetition pattern that is a square, as in Figure 2.18. Taking the larger square in the second tessellation to be of area equal to the octagon and taking the remaining square in that tessellation to be identical to the square in the first tessellation give two tessellations that we can superpose as in Figure 13.1. Lindgren (1951) published the first such superposition. The 4-fold rotational symmetry of each tessellation leads to 4-fold rotational symmetry in the dissection. Also, the dissection can be partially hinged.

How did Bennett come to discover such a wonderful dissection and dissection technique? Baker (1944) pointed out that Bennett's area of expertise was geometry and that Bennett was consulted by Percy MacMahon when MacMahon worked on his divisions of the plane. MacMahon (1921, 1922) described both of the tessellations in Figure 13.1. Thus Bennett's background was particularly suited to finding the beautiful 5-piece dissection.

In addition to Henry Dudeney's having been bettered by two pieces, there is another surprise associated with this dissection. For over half a century, someone other than Bennett has been given credit for it! And the circumstances are such that a case can be made for grand theft! Ball (1939), Lindgren (1964b), Cadwell (1966), and Gardner (1969) all have named James Travers as the originator. Travers states in the preface of his (1952) book, "I was the first, as Professor Coxeter agrees, in the revised edition of Raise [sic] Ball's *Mathematical Recreations* to dissect the regular octagon into the fewest pieces to form a square."

But Dudeney's attribution to Bennett predates the (Travers 1933) publication by seven years. Furthermore, in a letter of August 5, 1951, to Dick Gillings of Australia, Travers claims as his own George Wotherspoon's heptagon-to-square dissection, which had appeared in (Dudeney, *Strand*, 1927b). So it seems that Travers purloined not one but two dissections from the late columns of Henry Dudeney.

The dodecagon is similar to the octagon, with respect to completing the tessellation. As we saw in Figure 2.13, adding two equilateral triangles to the dodecagon forms a tessellation element. Also, a hexagon and two identical equilateral triangles form the tessellation element in Figure 2.16. In both cases the repetition pattern is that of a hexagon. If the tessellations of Figures 2.13 and 2.16 are superposed when the areas of the dodecagon and hexagon are equal and the triangles are all identical, then Figure 13.3 results. The associated 6-piece dissection is shown in Figure 13.4. Lindgren (1964b) credits this dissection to Ernest Irving Freese. It has 2-fold rotational symmetry and can be partially hinged. The four pieces other than the two small triangles can be hinged in one of two ways.

Geoffrey Thomas Bennett was born in London, England, in 1868. He distinguished himself in mathematics as a student at St. John's College, Cambridge, and upon completion of his studies was appointed college lecturer at Emmanuel College, Cambridge. He held a fellowship at the college from 1893 until his death in 1943.

Bennett's research specialty was the geometry of mechanism. In recognition of his work, he received a fellowship in the Royal Society in 1914 and a doctorate of science in 1920. During the First World War, he invented an altitude finder for antiaircraft guns and improved the design of a gyrocompass.

In his younger days Bennett was an exceptional bicyclist, reputed to be the fastest thing on the road. He had a machine constructed in which he lay flat along the frame, using a mirror to allow him to see forward. Bennett was a good pianist and at one time lectured on acoustics to the University Musical Society. His interests extended to experimenting with boomerangs and studying the structure of spider webs.

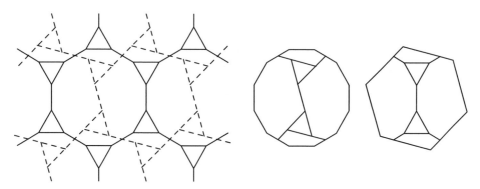

13.3: Tessellations: {12}, {6}   **13.4:** Dodecagon to hexagon

It is possible to form a tessellation from {6/2}s, if we add two equilateral triangles for each {6/2}, as shown in Figure 2.21. This is conveniently the same number
of triangles that we add to the hexagon to give a tessellation, as we have already
discussed. Thus we can superpose these tessellations to get a dissection of {6/2}
and {6}. Our goal is to get a solution in six pieces, as in Figure 13.4, in which there
are two equilateral triangles and four other pieces.

The difficulty is that the two equilateral triangles, when oriented as they would
be in the tessellation of {6}s and {3}s, cannot both fit completely inside the {6/2} at
the same time. The distance between the extremities of the two triangles is greater
than the width of the {6/2} at its narrowest point. Thus one equilateral triangle in
each tessellation must overlap some boundary in the other tessellation. If triangles
from each tessellation overlap, the area common to both need not be used in the
dissection. Positioning the tessellations as shown in Figure 13.5 gives the solution
in Figure 13.6. Each of the triangles that overlaps in the superposition of the tessellations is cut into three pieces, one of which is not used, and the other two of
which must be turned over in the dissection.

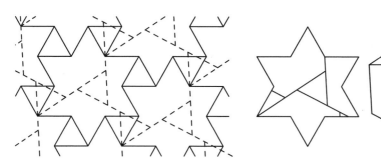

13.5: Tessellations: {6/2}, {6}   **13.6:** Hexagram to hexagon

Puzzle 13.1: Produce a 7-piece dissection in which one of the equilateral triangles is positioned in the center of each of the {6/2} and the {6}. How many pieces are turned over in this case?

My 6-piece solution (first reproduced in the second edition of (Gardner 1969)) improves on 7-piece solutions by Lindgren (1964b) and by Bruce Gilson in (Gardner 1969). However, each of those dissections has the advantages that no pieces are turned over and that there is no piece as small as the tiny piece in Figure 13.6.

The tessellation of hexagrams and triangles in Figure 13.5 appears to be unique in the following sense: There is no other tessellation containing a star and a regular polygon in which the star shares a common border of positive length with a neighboring star. Thus to find other examples of completing the tessellation that involve a star, we will add an irregular polygon to the star. We will then dissect this irregular polygon to the desired figure.

We find that this technique works for {8/3} to {4}. We can fit an {8/3} together with a 4-armed twisted cross to form a tessellation element. Serendipitously, this twisted cross is itself a tessellation element. Thus we dissect the 4-armed twisted cross to a square and use a square of this size, along with a square of area equal to that of the {8/3}, to form a third tessellation. The first and third tessellations are shown superposed in Figure 13.7, and the accompanying dissection is shown in Figure 13.8. This dissection has the same number of pieces as one by Lindgren

Ernest Irving Freese was born in Minnesota in 1886 and died in Los Angeles in 1957. For about forty-five years he practiced as an architect in Los Angeles, contributing to the design of buildings such as the Wilshire Country Club, the Southwest Museum, and the Southern California Automobile Club. He also planned homes and commercial buildings. During the First World War he designed concrete ships, and during the Second World War planned several army camps.

Freese published a 5-part series of articles on perspective projection and a 22-part series on the geometry of architectural drafting in *Pencil Points* during 1929–1932. Of peculiar charm is part 12, "Lines Akimbo," which Freese summarizes: "A Historical Romance appears upon the 'boards.' Geometrizing philosophers and philosophizing mathematicians climb down off their eternal pedestals in the Hall of Fame to tell us 'how'." Freese's "biography" appears in the March 1930 issue:

"**B**orn; **I**nformed **O**n **G**eometry;
Raised on **A**rchitectural **P**erspective;
Haint dead **Y**et!"

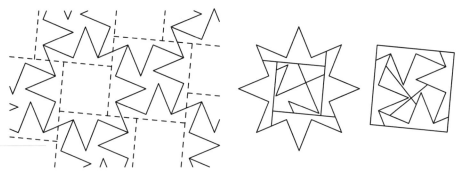

13.7: Tessellations: {8/3}, {4}   **13.8:** Octagram to square

(1964b), but it achieves greater symmetry. In particular, the four outer pieces of the star are translational and can be hinged. In a different orientation, the inner four pieces are translational.

We can also dissect a {12/4} to a square. First we form a tessellation element by combining a {12/4} with a ragged polygon that resembles a creature running, with its arms and legs flailing. This "ragged runner" is shaded in Figure 13.9. Although it is not itself a tessellation element, we can cut it into pieces to form a tessellation element with the repetition pattern of a square. Thus again the third tessellation is that of two squares, one of area equal to that of the {12/4}, and the other of area equal to that of the ragged runner. We superpose the first and third tessellations in Figure 13.9 but this time do not center them. Rather than cut the second tessellation element further, we cut niches in the third tessellation into which the element can fit. We leave the superposition uncentered in order to have sufficient room for the niches. The resulting 11-piece dissection is shown in Figure 13.10.

As a prelude to our next chapter, on Maltese Crosses, we dissect {M̆}, a version of the Maltese Cross designed by Bernard Lemaire. This version has a simple

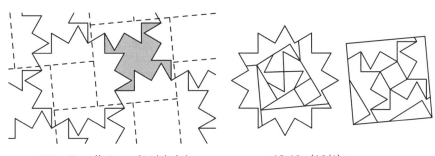

13.9: Tessellations: {12/4}, {4}   **13.10:** {12/4} to square

155

structure: All line segments forming its boundary are of equal length and have slopes whose absolute value is either 2 or 1/2. Shown in Figure 13.12 is my 5-piece dissection of {M̌} to a square, which appeared in (Berloquin, *Le Monde*, 1974b). Adding a small square to the cross gives a tessellation element. The second tessellation is the familiar tessellation of large and small squares. The superposition of these two tessellations is shown in Figure 13.11.

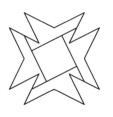

13.11: Tessellations: {M̌}, {4}   **13.12:** Cross to square

CHAPTER 14

Maltese Crosses

*The Great War had ended recently, and images of sol-
diers, slaughter, and valor were still fresh in people's
minds. For his "Perplexities" column in the April 1920
issue of* The Strand Magazine, *Henry E. Dudeney chose
the form of one medal, the Cross of Victoria, or Maltese
Cross, and posed the puzzle of dissecting it to a square
in the fewest possible number of pieces. His solution,
requiring thirteen pieces, followed in the May issue. As
memory of the war receded and wounds healed, the focus of everyday life moved on. And when, in 1926, collected puzzles from "Perplexities" appeared in* Modern
Puzzles and How to Solve Them, *a change had also occurred in the Maltese Cross puzzle. In place of the original solution was a beautiful new solution that used only
seven pieces! The remarkable dissection was attributed
by Dudeney to a Mr. A. E. Hill.*

In designing this pretty puzzle, Dudeney fixed the dimensions of the cross to conform to the 5×5 grid as shown in Figure 14.1. A. E. Hill's startling solution, given
in (Dudeney 1926a), is reproduced in Figure 14.2. Most attractive, it possesses 2-
fold rotational symmetry. However, a certain degree of mystery has accompanied
this dissection. First, no satisfactory explanation of the dissection method has been
given. Second, no information seems to be available on who A. E. Hill was. Although
I have had no luck in identifying Mr. Hill, I have found a reasonable explanation for
the dissection method.

Our explanation is suggested by the 13-piece dissection of Dudeney (*Strand*,
1920a), which is shown in Figure 14.3. Dudeney cut a wedge off each arm of the
Maltese Cross, leaving the outline of a square with four triangular cavities in it. He

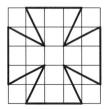

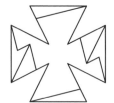

14.1: Maltese Cross   **14.2:** A. E. Hill's dissection

then cut each of the four wedges into three small pieces, which he rearranged to fill one of the triangular cavities.

How could A. E. Hill have found Figure 14.2, starting with Figure 14.3? We form Figure 14.4 by overlapping the cuts in the upper right part of the Maltese Cross and the square in Figures 14.2 and 14.3. These cuts partition the cross into regions that are identical to the small pieces in Figure 14.3. Each of the pieces in Figure 14.2, except the central piece, is the union of one each of the three small pieces from Figure 14.3. It would appear that A. E. Hill determined how to join the small pieces together so that moving the resulting pieces shuffled the small pieces in just the right way.

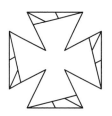

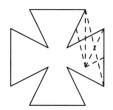

14.3: Dudeney's dissection   **14.4:** Small pieces

In fact, we can derive the pieces from a dissection by Gerwien (1833). We use the slightly simpler version by Fourrey (1907) in Figure 14.5. Triangle ABE has points

C and D on line segment BE such that line segments BC, CD, and DE are all of equal length. It follows that triangles ABC, ACD, and ADE are all of the same area. Then line segments are drawn from points C and D, parallel to AE, AD, AC, and AB. Each of triangles ABC, ACD, and ADE contains the same set of four pieces. The thicker outlines

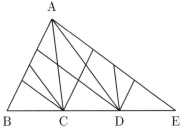

14.5: Gerwien's dissection

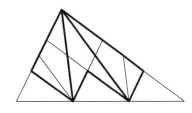

14.6: Three unions of pieces

in Figure 14.6 correspond to the three unions of pieces in Hill's dissection. Each of these is one of ABC, ACD, and ADE, minus the lowest piece. Thus the unions of pieces are all of equal area.

In going from the 13 pieces to 7 pieces, we lose some symmetry. Whereas the dissection in Figure 14.3 has 4-fold rotational symmetry, is translational, and can be fully hinged, the dissection in Figure 14.2 has only 2-fold rotational symmetry.

Puzzle 14.1: Modify Figure 14.2 to be translational, without increasing the number of pieces.

Puzzle 14.2: Where should the hinges be placed in a fully hinged model of Figure 14.3? Identify a problem if ordinary hinges are used on a completely flat model. Modify the basic hinge design to solve this problem.

What happens if we change Dudeney's dimensions for the Maltese Cross {M} but leave it in the same general form? We say that it is *well proportioned* if its area equals the area of the square defined on every second outer vertex. (These vertices are labeled A, B, C, and D in Figure 14.7.) An equivalent condition is that the distance between neighboring outer vertices equals the distance between opposite inner vertices (as shown with dotted lines in Figure 14.7).

Thus the cross in Figure 14.7 is well proportioned. We generalize the technique of Figure 14.2 to find the 7-piece dissection of this cross {M'} to a square in Figure 14.8. A dissection corresponding to that in Figure 14.3 would use 21 pieces: a central piece, plus four sets of 5 pieces, where each set converts one wedge from the Maltese Cross so that it fills a triangular cavity in the outline of the square. In fact, there is a 7-piece solution whenever the underlying grid is of the form $(2k + 1) \times (2k + 1)$, for any integer $k > 1$.

159

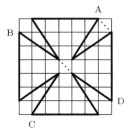
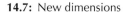

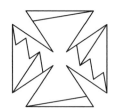

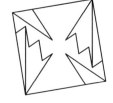

14.7: New dimensions

14.8: A new dissection

Puzzle 14.3: Generalize the dissection in Figure 14.8 to give a 7-piece dissection for a well-proportioned cross on a 9×9 grid.

Puzzle 14.4: What figure results from generating a well-proportioned cross on a 3×3 grid? What dissection of the resulting figure will correspond to the dissection in Figure 14.3?

Before learning of Hill's 7-piece solution, Lindgren (1961) gave an 8-piece dissection. His dissection (Figure 14.10) resulted from superposing a tessellation of squares over the tessellation of the element in Figure 14.9. It works for any well-proportioned cross, not just those that fit on a grid, and has 2-fold rotational symmetry.

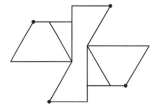

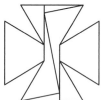

14.9: Lindgren's element

14.10: Lindgren's dissection

Puzzle 14.5: Lindgren's dissection simplifies to a 4-piece solution for the cross on a 3×3 grid. Give the solution in this case.

Puzzle 14.6: Lindgren's dissection simplifies to a 6-piece dissection for a well-proportioned cross on one grid that is not 3×3. Determine which grid allows a 6-piece solution, and give the dissection.

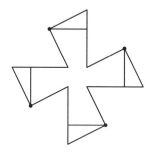

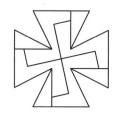

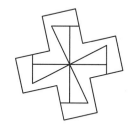

14.11: Lemaire's element   **14.12:** Lemaire's Maltese to Greek

Since Lindgren's tessellation element for the Maltese Cross (in Figure 14.9) has the repetition pattern of a square, as does a Greek Cross, an economical dissection of the Maltese Cross and Greek Cross is possible. Lindgren (1961) gave such a dissection, using nine pieces. Bernard Lemaire found a different tessellation element for the Maltese Cross, with four pieces turned over, presented in an equivalent form in Figure 14.11. Lemaire's 8-piece dissection of the Maltese Cross and Greek Cross, appearing in (Berloquin, *Le Monde*, 1975c), is shown in Figure 14.12. The result of superposing the standard tessellation for the Greek Cross with Lemaire's tessellation, this beautiful dissection possesses 4-fold rotational symmetry. Lemaire's dissection works for any well-proportioned cross.

To dissect a Maltese Cross to a Greek Cross in eight pieces, is it necessary that some pieces be turned over? Anton Hanegraaf found a tessellation element for the Maltese Cross {M} that turns no pieces over, as shown in Figure 14.13.

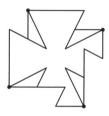

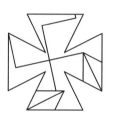

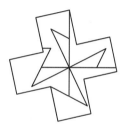

14.13: Hanegraaf's element   **14.14:** Hanegraaf's Maltese to Greek

Superposing the tessellation for this element with the tessellation of Greek Crosses gives the 8-piece dissection in Figure 14.14. In addition to turning no pieces over, this dissection is translational. However, it does not appear to generalize to other well-proportioned crosses.

Lemaire discovered other dissections for other crosses that are not well proportioned. An example is the cross {M'''} outlined in Figure 14.15. Each diagonal side of an arm of the cross is half the length of the side of an equivalent square. This makes possible Lemaire's clever 8-piece dissection, published in (Berloquin, *Le Monde*, 1975a), and shown in Figure 14.16. Although four pieces are turned over in this dissection, it has 4-fold rotational symmetry.

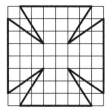 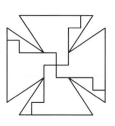 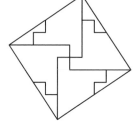

14.15: Lemaire's cross   **14.16:** Lemaire's cross to square

Bernard François Lemaire was born in Paris in 1945. He completed studies in electrical engineering at the École Supérieure d'Électricité in 1968, where he specialized in computer science and operations research. At the Université Pierre et Marie Curie (Paris VI), he received a doctorate in operations research in 1971 and a Thèse d'État in 1978. He joined the Conservatoire National des Arts et Métiers in Paris in 1968 as a first lecturer in computer science; became senior lecturer in operations research in 1972, then assistant professor in 1978; and has held the operations research chair since 1981. His research interests include stochastic processes and Petri nets.

In recreational mathematics, Bernard has dissected all sorts of crosses, and has solved problems on $n \times m$ chessboards, including knight "covers," knight "attacks," and uncrossed knight rides. He enjoys languages (English, German, and Russian), as well as mountain hiking and camping. He aspires to being handy around the house but has been told by his wife that he's better at cutting crosses!

CHAPTER 15

Curves Ahead

My eyes grew rounder as I rolled the microform image forward on the screen. The wording of both the puzzle and its solution differed just enough from the later versions in books to raise a doubt. Perhaps Sam Loyd had not discovered the 6-piece solution to the oval-stool-top puzzle. Perhaps it was sent in by a reader.

Ninety years before, had Loyd's eyes also grown rounder as he examined the reader's letter? Had he been startled to see an improved solution to a puzzle that had become an old saw? Had he wondered what John Jackson, the originator of the puzzle, might have wondered eighty years before? Had Loyd imagined that some day the puzzle might turn into the puzzle of the puzzle, and in a curious way be brought full circle?

John Jackson (1821) posed what may be the earliest published dissection of curved figures. He converted a circular table top (a disk) to the seats of two oval stools with open handholds. His simple 8-piece dissection is shown in Figure 15.1. It has a circular cut centered on, and half the diameter of, the original disk. Two straight cuts, of diameters perpendicular to each other, complete the dissection. Half of the four inner and half of the four outer pieces reassemble to make each stool seat.

Jackson's dissection is not minimal, since Loyd (*Inquirer*, 1901a) gave the 6-piece dissection in Figure 15.2. The disk is first cut into the two shapes that form the Chinese monad (or Yin and Yang symbol). Then the tails of the Yin and Yang pieces are cut off to fill in the portion of the outline of the seat top claimed by the other of the Yin and Yang pieces. Finally, two pieces are cut out to reveal the handholds and then used to complete the outline of the seat tops.

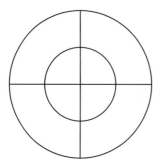 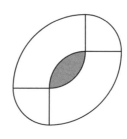 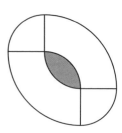

15.1: Jackson's disk to oval seat tops

John Jackson was a "Private Teacher of the Mathematics," who lived in England from 1793 to 1875. He was the author of a puzzle book containing mathematical, geographical, and scientific recreations.

So why the suspicion that Loyd appropriated this dissection from a reader? In his *Philadelphia Inquirer* columns, Loyd left open his dissection puzzles, asking only for a dissection in the fewest number of pieces. He encouraged readers to send in solutions and awarded prizes for the best. We know of no instance in which Loyd acknowledged that a reader had found a better solution. Could he have been that infallible? Loyd was careful to emphasize that he took the puzzle from the old puzzle books and then reversed it because it was otherwise viewed as too difficult. It makes sense that this reversal is what he meant by "new teeth." When Loyd gave the 6-piece solution three weeks later, he first supplied the old 8-piece dissection and then the new dissection and said that from then on the 6-piece dissection must be taken as the solution for the puzzle. If he had known of the 6-piece solution when composing the puzzle, would he not have emphasized that the solution in the old puzzle books was not the best possible?

In this chapter we broaden our scope of figures to include those that have curved edges. Dissecting curved figures is tricky. How do we choose which figures to dissect and what their precise shape should be? The circle is a natural choice, but the oval seat top is certainly not universal. How should a crescent or a heart be formed? Also, to fit a curved side of one piece flush with a curved side of another piece, the curvature of the boundaries must match. These issues conspire to make the dissection of curved figures an exercise in searching for suitable pairs of figures. For this reason, we will confine our survey primarily to older puzzles that have historical interest. These puzzles, such as a crescent to a cross, and a spade to a heart, have undeniable charm.

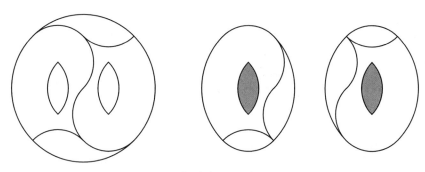

15.2: Loyd's disk to oval seat tops

Loyd's dissection of the disk and seat tops does not retain all the symmetry in Jackson's dissection, such as translation with no rotation.

Puzzle 15.1: Modify Jackson's disk to oval seat tops dissection to get a 6-piece dissection that is translational.

Both Loyd and Dudeney posed variations on the oval seat top puzzle. Dudeney (*Dispatch*, 1902d) made the ovals more pointed, and the handholds consequently narrower. His variation still used just six pieces. Loyd left the oval seat tops with the same outline but rotated the handholds 90°. Dudeney (*Strand*, 1927a) attributed the pleasing 4-piece dissection in Figure 15.3 to Loyd.

The competition between Loyd and Dudeney extended to the puzzle of dissecting a crescent to a Greek Cross. Each had his own form for the crescent. True to Loyd's advertising instincts, the graphic for the puzzle, shown on the next page, mixes the images of a beautiful woman, a lustrous Greek Cross, and a bright crescent moon. The Cross and Crescent Puzzle first appeared in (Loyd, *Press*, 1900a).

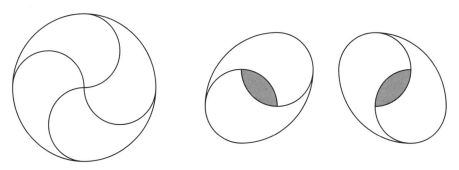

15.3: Loyd's disk to different oval seat tops

165

Loyd's crescent, which we designate by {C}, has two matching circular arcs, with line segments of equal length joining the ends of the arcs. The line segments appear to be separated by a distance that is five times their length, and the radius of the arcs appears to be $\sqrt{26}/2$ times the length of the line segments. Lindgren (1964b) pointed out that Loyd's dissection does not seem to be exact. However, we give a similar, exact 7-piece dissection in Figure 15.5.

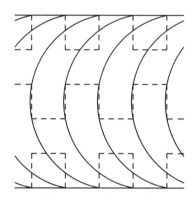

15.4: Suggestive overlap   **15.5:** Crescent to cross

To discover the dissection, nestle crescents inside one another, then superpose a strip that contains an unusual amalgam of Greek Crosses, as shown in Figure 15.4. The inferred cuts for one crescent are the reflection of the cuts for the next crescent. One of the pieces must be turned over.

Loyd's crescent is well suited for dissection to a Greek Cross since the crescent's height is five times its width and the Greek Cross consists of five squares. More challenging is the crescent's dissection to a square. Once again, we nestle the crescents together. Now, however, we overlay them with a sequence of figures shown with dashed edges in Figure 15.6. The figures consist of two squares, whose side lengths are in the ratio of 1 : 2, that are attached and are of combined area equal to that of the crescent. The dotted edges in the bottom middle figure indicate a standard dissection that converts the figure to a square. We have overlaid the sequence of figures so that the dotted cuts cross as few edges as possible. The resulting 8-piece dissection is in Figure 15.7. Half of the pieces must be turned over.

Dudeney (*Dispatch*, 1899a) gave another shape for the crescent. It has two semicircular arcs, with line segments of equal length joining the ends of the arcs. The line segments are separated by a distance that is four times their length, so that the radius of each semicircle is twice the length of a line segment. This crescent {C'} is less

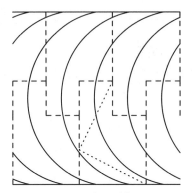

15.6: Overlay of figures   **15.7:** Crescent to square

suggestive of a true crescent than Loyd's. Dudeney (*Dispatch*, 1899a) gave a 4-piece dissection of the crescent to a square. Using ten pieces, Dudeney (*London*, 1902a) gave a three-way dissection of the crescent to a square, and then to a Greek Cross. Dudeney (1907) stated that he had a dissection of the crescent directly to a Greek Cross "in considerably fewer pieces, but it is far more difficult to understand than the above method, in which the problem is simplified by introducing the intermediate square." I have not been able to locate this dissection.

However, Lindgren (1964b) found a 7-piece dissection. He overlaid a Greek Cross strip with a crescent strip that he produced by cutting the crescent in half and then turning over one of the two pieces. The overlaid strips are shown in Figure 15.8, and the resulting dissection is in Figure 15.9. Could this have been the dissection that Dudeney referred to? Probably not, because we can derive this dissection by the same method Dudeney employed in his 10-piece dissection: In Figure 15.8, the dotted lines show the overlay of squares, yielding the same 4-piece dissection of crescent to square. The implied square to Greek Cross dissection is the same as the one Dudeney used, but with the pieces rearranged in the square.

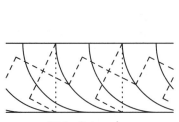

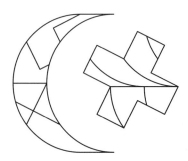

15.8: Stripped up   **15.9:** Crescent to cross

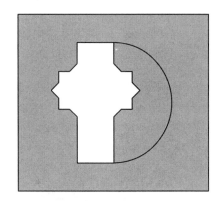

15.10: Loyd's flags: star and crescent to Crusader's Cross

Loyd (*Inquirer*, 1898a) used the same crescent as in Figure 15.9 in another dissection. In the Crusader's Prize Puzzle, he composed a flag from a dark piece of cloth sewn over a light piece. He cut a crescent and an {8/2} out of the dark piece, revealing the light piece below. Loyd asked his readers to cut the dark piece and rearrange the resulting pieces to leave a hole in the form of a scepter-shaped cross. The dissection requires only two pieces, as shown in Figure 15.10. We call this type of dissection a *hole dissection*, because the objects of interest, that is, the star, the crescent, and the scepter, are actually holes in the surface that we intend to cut.

Loyd (*Inquirer*, 1901c) also presented a remarkable curiosity, a 3-piece dissection of a spade to a heart in the Puzzle of the Red Spade:

> During a recent visit to the Crescent City Whist and Chess Club my attention was called to the curious feature of a red spade which appears in one of the windows of the main reception room.... No reason was ever vouchsafed, nor even asked for, regarding the incongruousness of the color of the emblem. It was looked upon as a blunder which occasioned comment at first, but which came to be looked upon afterwards with favor, not only on account of the novelty of such a thing as a red spade, but from the recognized point that a black spade would make the room too dark. Hearing accidentally, however, that a blunder had actually been committed by the manufacturer, in that the ace of hearts was to have been the insignia of the club, I was led to examine the window carefully and found that the spade was composed of three pieces, and speedily discovered that by proper arrangement of the three pieces they would be fitted together so as to form the ace of hearts, as originally desired.

As shown in Figure 15.11, Loyd cut out a wedge that contains the tail of the spade; separated the two remaining pieces; and reinserted the wedge, to give a heart.

169

PROPOSITION—Show how to change a spade into a heart.

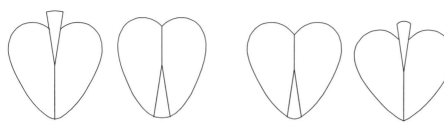

15.11: Spade to heart   **15.12:** Heart of darkness?

Yet the shape of the resulting heart is a bit odd: Has the human heart changed so much in the last one hundred years? Such aesthetic issues underscore the problem that we have faced in this chapter: How should one choose an appropriate model for a curved figure such as a crescent or a heart? Strangely, Loyd's model of the spade would do much better for the heart, if the tail of the spade were removed. Trying this and working backward to get a spade gives the form and consequent dissection in Figure 15.12. These new forms of both heart and spade look acceptable, prompting speculation: Could the puzzle have started off as converting a "black heart," which for editorial reasons then became converting a red spade?

CHAPTER 16

Stardom

Harry Lindgren looked up to survey the evening sky: Those millions of points of light were not the ones of his youth. For he had left a homeland of dimmed prospects to seek opportunity "down under." Each blinking light, which long ago had seemed so small and twinkly, he now knew to represent an object of magnificent size and remarkable structure. And if he were to have surveyed his own life, he might have found a curious resonance.

Denied the chance for postgraduate study in mathematics, he had focused on mathematical recreations. His work on geometric dissections had supernovaed, and he had come to outshine previous luminaries, Henry Dudeney and Sam Loyd included. And when his book was almost finished, he had made a discovery that would ensure his place among the stars. He had found the key to their structure, and with it an elegant new type of dissection, based on simple trigonometric relationships. For us then, he left another sky – his own – full of those beautiful star dissections.

Harry Lindgren's constellations of star dissections take advantage of, and celebrate, the internal structure of polygons and stars first discussed in Chapter 2. He pioneered dissections among pairs of stars and polygons, such that cuts need only be along the sides or diagonals of the constituent rhombuses. Although Geoffrey Mott-Smith (1946) was the first to take advantage of this structure in his dissection of a hexagram to a triangle (Figure 10.21), it was Lindgren who recognized the general principle and used it to create a wealth of what we term *auspicious* dissections.

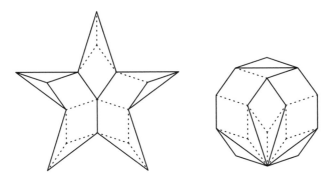

16.1: Pentagram to decagon

A masterful example by Lindgren (1964b) is shown in Figure 16.1: a 6-piece dissection of a pentagram to a decagon. Recall from Chapter 2 that the decagon consists of five 72°-rhombuses and five 36°-rhombuses, indicated by dotted lines in the decagon of Figure 16.1. Remarkably, one of the internal structures of the pentagram consists of the same set of rhombuses, also indicated with dotted lines. Lindgren split the 36°-rhombuses lengthwise on their diagonals in the pentagon to give the side of the pentagram. Thus the dissection relies on the relationship, $s\{5/2\}/s\{10\} = 2\cos(\pi/10)$, between the side lengths of the pentagram and the decagon. Harry Lindgren catalogued a large number of these trigonometric relationships.

An equally masterful dissection, of one $\{10/3\}$ to two $\{5/2\}$s, is shown in Figure 16.2. One internal structure of a $\{10/3\}$ consists of ten 72°-rhombuses and ten 36°-rhombuses, as indicated by dotted lines in Figure 16.2. The ratio $r\{10/3\}/s\{5/2\} = \sqrt{2}$ between the short radius of a $\{10/3\}$ and the side of a pentagram of equal area makes this dissection possible. Besides possessing rotational symmetry, the dissection can be partially hinged, as in Figure 16.3. The underlying

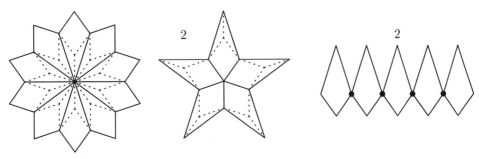

16.2: $\{10/3\}$ to two $\{5/2\}$s

16.3: Hinged $\{5/2\}$s

173

relationship also applies to other sets of figures, such as two {3}s to one {6}, and two {7/3}s to one {14/5}.

Lindgren expressed the trigonometric relationships of stars and polygons in terms of a few basic quantities. For polygons, these are s, the side length of the polygon, and R, the length of the radius (from the center to a vertex). For stars, we measure the side length s from a vertex to an adjacent reflex angle. There is a second radius r for stars, the distance from the center to a reflex angle. We identify these dimensions for a {5/2} and a {10} in Figure 16.4.

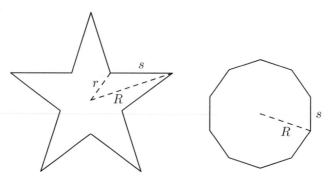

16.4: Sides and radii of polygons and stars

Let's explore the "galaxies" first charted by Lindgren (1964b). For each, Lindgren discovered a trigonometric relationship that determines it. These relationships hold for every positive integer n.

$$\frac{s\{n+2\}}{s\{(2n+4)/2\}} = 4\cos(\pi/(2n+4)) \qquad (16.1)$$

$$\frac{R\{2n+2\}}{s\{(2n+2)/n\}} = \sqrt{2} \qquad (16.2)$$

$$\frac{s\{(2n+2)/n\}}{s\{(4n+4)/2\}} = \sin(\pi/4)/\sin(\pi/(4n+4)) \qquad (16.3)$$

$$\frac{s\{4n+2\}}{s\{(4n+2)/(n+1)\}} = \sqrt{2} \qquad (16.4)$$

$$\frac{s\{(4n+2)/(n+1)\}}{r\{(4n+2)/(2n)\}} = 2\sin(\pi/(4n+2)) \qquad (16.5)$$

$$\frac{s\{(4n+2)/(n+1)\}}{s\{(8n+4)/2\}} = 4\cos(\pi/4)\cos(\pi/(8n+4)) \qquad (16.6)$$

$$\frac{r\{(4n+2)/(2n-1)\}}{s\{(2n+1)/n\}} = \sqrt{2} \tag{16.7}$$

$$\frac{R\{(8n-2)/(2n)\}}{R\{(4n-1)/n\}} = \sqrt{2}\cos(\pi n/(4n-1)) \tag{16.8}$$

$$\frac{s\{(6n+2)/(2n+1)\}}{s\{(6n+2)/n\}} = 2\cos(\pi n/(6n+2)) \tag{16.9}$$

$$\frac{s\{(6n+3)/(n+1)\}}{s\{(6n+3)/(3n+1)\}} = 2\cos(\pi(3n+1)/(6n+3)) \tag{16.10}$$

$$\frac{s\{(6n+3)/(n+1)\}}{r\{(12n+6)/(6n+1)\}} = \sqrt{2}\cos(\pi(3n+1)/(6n+3)) \tag{16.11}$$

Lindgren (1964b) in Chapter 20 and Frederickson (1972a) explained how the relationships are derived. We must first find expressions for the area of polygons and stars in terms of s, r, and R. For example, to derive relationship (16.1), we can obtain expressions for the areas of the polygon $\{n+2\}$ and the star $\{(2n+4)/2\}$:

$$A\{n+2\} = \frac{(n+2)s^2\{n+2\}}{4\tan(\pi/(n+2))}$$

$$A\{(2n+4)/2\} = \frac{(2n+4)\cos(\pi/(2n+4))\cos(2\pi/(2n+4))s^2\{(2n+4)/2\}}{\sin(\pi/(2n+4))}$$

Setting these expressions equal, and using the identity $\sin(2x) = 2\sin(x)\cos(x)$ with $x = \pi/(2n+4)$, gives

$$\frac{s^2\{n+2\}}{s^2\{(2n+4)/2\}} = 16\cos^2(\pi/(2n+4))$$

Taking the square root of both sides gives relationship (16.1). When the denominator in the notation for a star equals 1, as in relationship (16.2) with $n = 1$, we should interpret the star as a polygon. In this case, however, the value of the right-hand side differs from the stated value, though a special relationship still exists.

Galaxy (16.1), with $n = 1, 2, 3, 4$, contains four dissections in Chapter 10: $\{6/2\}$ to $\{3\}$ in Figure 10.21, $\{8/2\}$ to $\{4\}$ in Figure 10.15, $\{10/2\}$ to $\{5\}$ in Figure 10.25, and $\{12/2\}$ to $\{6\}$ in Figure 10.19. These use five, seven, seven, and eight pieces, respectively. We have already asked, Could there be a 6-piece dissection for $\{8/2\}$ to $\{4\}$? Another interesting 7-piece dissection for $\{8/2\}$ to $\{4\}$ is given in Figure 16.5. I invite our octonauts to search for the elusive 6-piece dissection, if it exists.

I have explored this galaxy more extensively. For all odd values of n in (16.1), I described in (1972a) a $(2n+4)$-piece dissection of $\{n+2\}$ to $\{(2n+4)/2\}$. But we already know that there are more economical dissections for $n = 1$ and $n = 3$. In

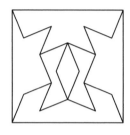

16.5: {8/2} to square

addition, I have found an 11-piece dissection of a {14/2} to a {7}. Since the first three cases have better dissections, is there a better general method?

For all even values of n, Elliott (1983) blazed forth with a $(3n+6)$-piece dissection of {$n+2$} to {$(2n+4)/2$}. But I have found a $(2n+4)$-piece dissection for all even values of $n > 4$. Figure 16.6 illustrates the case $n = 6$, which we base on Elliott's dissection. The dotted lines hint at how to derive some of the pieces in our dissection from Elliott's simpler dissection by trading isosceles triangles between pieces. Although our dissection has fewer pieces than Elliott's, $(n+2)/2$ of them must be turned over. When $n + 2 = 4$ and $n + 2 = 6$, we have $(n+5)$-piece and $(n+4)$-piece dissections, respectively. Can a reader find a general approach that improves on the $(2n+4)$-piece bound?

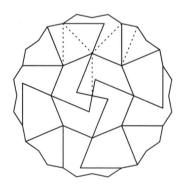
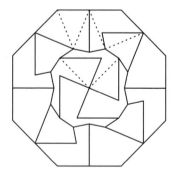

16.6: {16/2} to octagon

Puzzle 16.1: Produce the (nonoptimal) dissection for $n = 4$ that is analogous to the inferred dotted dissection in Figure 16.6.

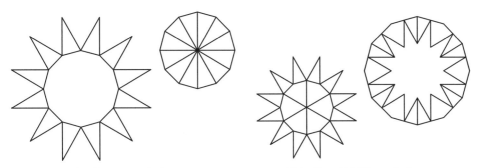

16.7: {12/5} to two dodecagons   **16.8:** Two {12/5}s to dodecagon

Are there "binary star" systems in Lindgren's universe? When the right-hand side of a relationship equals $\sqrt{2}$, then there should be a simple dissection of two figures into one. Thus galaxy (16.2) holds simple dissections of two $\{2n+2\}$s to $\{(2n+2)/n\}$, and two $\{(2n+2)/n\}$s to $\{2n+2\}$. Lindgren identified $(2n+3)$-piece and $(3n+4)$-piece dissections, respectively, shown for $n = 5$ in Figures 16.7 and 16.8. The bases for the internal structures are (π/n)-rhombuses and halves of such rhombuses. We can partially hinge the first dissection. We can also variously and partially hinge the second, if we allow a different arrangement of the pieces in the large {12}, namely, that all pieces from the center of the dissected {12/5} be consecutive around the outside of the large {12}.

But galaxy (16.2) has a second surprise. When n is odd, each figure has a number of vertices that is a multiple of 4. Since $\sqrt{2} = 2\sin\pi/4$, simple two-to-one dissections are possible in this case. For $n = 3$, Lindgren gave a 6-piece dissection of a {8/3} to octagon (Figure 16.9). The dotted lines indicate the structure of the star and the octagon, which consists of eight halves of 45°-rhombuses. Lindgren further subdivided such a half-rhombus into an isosceles right triangle and a smaller right triangle.

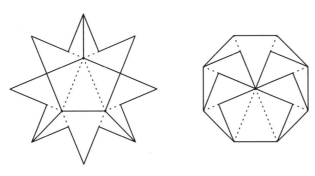

16.9: {8/3} to octagon

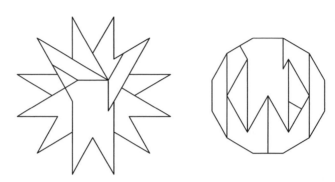

16.10: {12/5} to dodecagon

For $n = 5$, Lindgren gave an 11-piece dissection of a {12/5} to a dodecagon, which I improved to ten pieces in (1972a). The improvement involves cutting portions off certain pieces and reattaching them to other pieces, an approach that we have already seen in Figure 16.6. The dissection is shown in Figure 16.10.

One of the more "startling" of the galaxies is (16.3), which has auspicious dissections of {(4n+4)/2} to {(2n+2)/n}. Lindgren gave several 9-piece dissections for the case when $n = 2$. We see a variant of one of these in Figure 16.11. Sometimes galaxies seem to overlap, because the figure also illustrates the case of $n = 1$ for (16.6).

Galaxy (16.4) has auspicious dissections for two {4n+2}s to {(4n+2)/(n+1)}, and for two {(4n+2)/(n+1)}s to one. Lindgren identified (2n+2)-piece and (4n+2)-piece dissections, respectively. These should already be familiar to our apprentice stargazers, since they appear within several other dissections. For $n = 1$, the dissection by Lindgren of two {6}s to one {6/2} appears as part of our Figure 6.16, in the center of the large hexagon. Also, the dissection by Lindgren of two {6/2}s to one

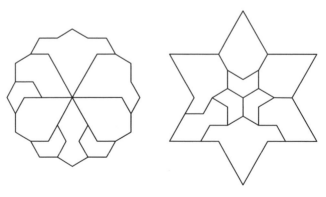

16.11: {12/2} to {6/2}

{6} appears as part of our Figure 17.8, where two hexagrams fill in the hexagonal center of the larger {6/2}.

For $n = 2$, the dissection by Lindgren of two {10}s to {10/3} appears in solid lines in Figure 22.9. For general n, there are $2n + 2$ pieces. Lindgren gave two different 10-piece dissections for two {10/3}s to {10}. We find a variation of one of them, which is translational, in Figure 16.12. For general n, this has $4n + 2$ pieces. As in Figure 16.12, there is always a dissection such that the {4n+2} contains a "galactic arm" of $2n$ rhombuses going from one side to another through its center.

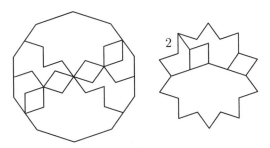

16.12: Two {10/3}s to one {10}

Let's investigate some stars on a "main sequence" of star dissection. Galaxy (16.5) has auspicious dissections of a {(4n + 2) / (2n − 1)} to a {(4n + 2) / (2n)}. Both expressions evaluate to {6/2} when $n = 1$, but for $n = 2$, we get a {10/3} to a {10/4}. Lindgren gave an 11-piece dissection, which is shown in Figure 16.13, where we indicate an internal structure by the dotted lines.

Galaxy (16.7) has auspicious dissections of two {(4n + 2)/(n + 1)}s to one {(2n + 1)/n}, and vice versa. For the case of $n = 1$, Lindgren gave what appears to be the only 4-piece dissection of one {6} to two {3}s. We dissected the smaller

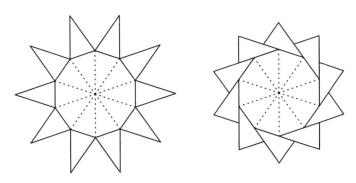

16.13: {10/4} to {10/3}

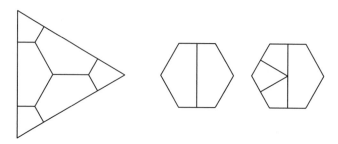

16.14: Two hexagons to one triangle

hexagon in Figure 5.16 this way. For two {6}s to one {3}, Lindgren gave a 6-piece dissection. In Figure 16.14, we see an alternate 6-piece dissection taken from (Paterson 1989). When $n = 2$, we have already seen a dissection of two {5/2}s to one {10/3} in Figure 16.2.

Increasing our speed to warp-7, we head for galaxy (16.8). When $n = 2$, I discovered in (1972a) an 18-piece dissection for two {7/2}s to one {14/3}. We now zoom past a 3-piece improvement (Figure 16.15). The dissection is translational and comes within a whisker, both figuratively and literally, of having reflection symmetry.

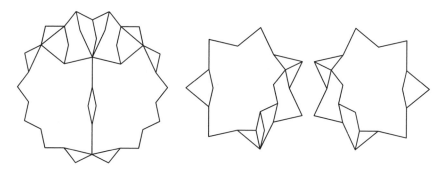

16.15: One {14/3} to two {7/2}s

Boosting our speed to warp-9, we head for galaxy (16.10). Our mathematics officer has targeted the case for $n = 1$. In (1972a) I gave a 14-piece dissection of a {9/2} to a {9/4}. Subsequently, in (1974), I found an 11-piece dissection. Shown in Figure 16.16, this dissection will be hard to beat. One piece contains three of the points of the {9/4}, and we nestle the remaining six points of the {9/4} into its two large cavities when forming the {9/2}. To accommodate the remaining pieces, we cut a notched piece out of another piece in the {9/2}. Not only does this make room

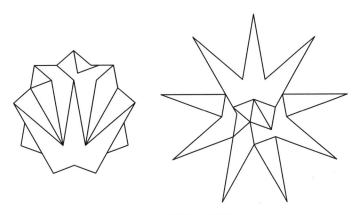

16.16: {9/2} to {9/4}

for the two rhombus halves to nestle in the {9/4}, but it creates a piece that exactly fills an awkward space in the {9/4}.

Besides relationships that determine galaxies of dissections, Lindgren (1964b) and Frederickson (1972a) identified relationships that seem to be part of no general pattern. These give rise to some spectacular starlike voyagers, or "comets."

$$\frac{s\{5/2\}}{s\{10\}} = 2\cos(\pi/10) \tag{16.12}$$

$$\frac{r\{10/4\}}{r\{5/2\}} = 2\cos(\pi/5) \tag{16.13}$$

$$\frac{s\{4\}}{R\{12\}} = 2\cos(\pi/6) \tag{16.14}$$

$$\frac{s\{4\}}{s\{12/5\}} = 2\cos(\pi/12) + 2\sin(\pi/12) \tag{16.15}$$

$$\frac{s\{3\}}{s\{12/2\}} = 6 + 4\cos(\pi/6) \tag{16.16}$$

$$\frac{s\{7/3\}}{s\{14/2\}} = \sqrt{2}\cos(3\pi/14)/\sin(\pi/14) \tag{16.17}$$

$$\frac{s\{7/3\}}{s\{7\}} = \sqrt{2} \cdot 2\cos(\pi/7) \tag{16.18}$$

$$\frac{s\{7\}}{s\{14/5\}} = \cos(\pi/14)/\cos(\pi/7) \tag{16.19}$$

$$\frac{s\{6\}}{s\{9/3\}} = 2\cos(\pi/18) \tag{16.20}$$

Since these comets are both fundamental and beautiful, it is not surprising that we have already spotted some of them elsewhere in the book. Harry Lindgren's 6-piece dissection of a {5/2} to a {10} was the pole star for this chapter, and it shall guide us again in Figures 16.21–16.23. Next, Lindgren gave an 18-piece dissection of two {5/2}s to one {10/4}, which we reproduce as level 2 of the prism dissection in Figure 22.8. For {12} to {4}, we have already seen one of Lindgren's 6-piece dissections in Figure 10.5.

For {12/5} to {4}, we need no telescopes. Lindgren gave two lustrous 10-piece dissections. In (1972a) I improved one of the two dissections by cutting sections off certain pieces and reattaching them to others. My dissection, shown in Figure 16.17, does not possess the nice 2-fold rotational symmetry of the other of Lindgren's dissections but uses only nine pieces.

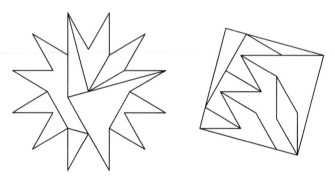

16.17: {12/5} to square

Of unparalleled brilliance is Lindgren's remarkable 6-piece dissection of a {12/2} to a triangle. With its elegant reflection symmetry, it is one of the few dissections that rival the octagon to square dissection of Figure 13.2. We gaze upon a variation of Lindgren's dissection in Figure 16.18.

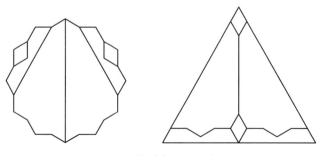

16.18: {12/2} to triangle

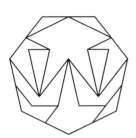

16.19: Two {7/3}s to one heptagon

We now move on to "comets with seven tails." For two {7}s to one {7/3}, we have already seen a 9-piece dissection in Figure 1.1. For two {7/3}s to one {7}, I gave a 14-piece dissection in (1972a). However, I have since reduced the number of pieces to 12, in Figure 16.19.

Do we dare hope for a "9-tailed comet"? In (1972a) I gave a 9-piece dissection of a {9/3} to a hexagon. Subsequently in (1974), I improved this to eight pieces, with one piece turned over. This latter dissection is shown in Figure 16.20.

Now we have reached the boundary of Lindgren's universe and should be alert for startling new phenomena. But are we prepared to discover that we can base Lindgren's {5/2} to {10} dissection on a certain type of tessellation? In Chapter 10 we extended plane tessellations to polyhedral tessellations, then extended polyhedral

Harry Lindgren was born in Newcastle-on-Tyne, England, in 1912. After his schooling, he completed an electrical engineering draftsman apprenticeship but could find no work, because of the Great Depression. He emigrated to Perth, Australia, in 1935 and earned a bachelor of science and education diploma at the University of Western Australia. He taught school in Western Australia for several years before moving in 1946 to Canberra, where he worked as a patent examiner in electrical specifications until retiring in 1972.

Harry established himself as the world's expert in geometric dissections, with his discoveries appearing in a number of journals and ultimately in his book *Geometric Dissections* in 1964. Shortly thereafter, his interests shifted to English language spelling reform, for which he championed a gradual change to a phonetic system. He published the book *Spelling Reform: A New Approach* in 1969, and also put out a quarterly newsletter, *Spelling Action*.

Harry played the violin and was a member of the Canberra Orchestra. He found pleasure in mastering several languages. He enjoyed reading books and drinking tea, which he made with great ceremony. He died in 1992.

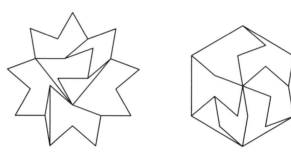

16.20: {9/3} to hexagon

tessellations to include dihedral tessellations, allowing polyhedra that contain volume of measure 0. We tolerated these latter polyhedra because we were concerned not with the volume inside the polyhedron, but only with its surface.

But why not venture a light-year further and allow surfaces that do not enclose volume because they are "nonorientable"? Examples of such surfaces are the "projective plane" and the "Klein bottle." These surfaces cannot be constructed in three dimensions, but this limitation does not matter for our purposes. By superposing tessellations on such surfaces, we can derive two dissections, including Lindgren's beautiful {5/2} to {10} in Figure 16.1.

We start with a decagon in Figure 16.21 and make cuts along three line segments to give a tessellation element. Ten of these tessellation elements cover a nonorientable surface formed by gluing together pairs of edges from ten pentagrams. As with our previous polyhedral dissections, we fold our tessellation element, with the folds corresponding to edges of the desired figure, in this case, a pentagram. The folds are shown as dotted lines in Figure 16.21 and numbered 1 through 5. One face of the nonorientable surface is shown in Figure 16.22, with the edges numbered to correspond to the folds. The gluing is consistent with the numbering,

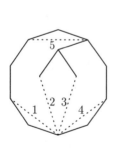

16.21: Element for {10}

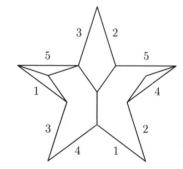

16.22: Edge pairs in {5/2}

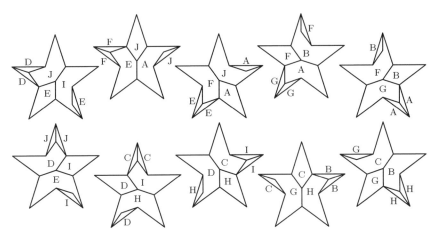

16.23: Superposition of nonpolyhedral tessellations: {5/2}, {10}

so that we glue a pair of like-numbered edges in one pentagram to a pair of like-numbered edges in another pentagram, with the outside points matched up.

The gluing operation forces the use of ten pentagrams, corresponding to the ten combinations of rotation and reflection that map a pentagram onto itself. The ten pentagram faces of the nonorientable surface are shown in Figure 16.23. Half of the faces are a rotation of Figure 16.22, and the other half are a reflection of a rotation. We arrange them in a symmetric manner, taking the first row from left to right, followed by the second row from right to left. We label the pieces in each face with letters from A to J, with each letter labeling all pieces originating from the tessellation element of one of the ten decagons.

We also derive a variant of our two {7}s to {7/3} by using nonorientable surfaces for tessellations. Starting with two heptagons, we cut an irregularly shaped

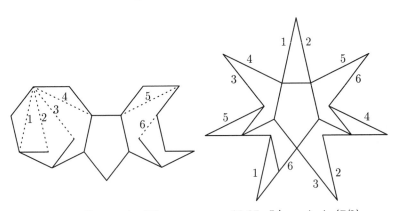

16.24: Element for {7}s   **16.25:** Edge pairs in {7/3}

185

16.26: Superposition of nonpolyhedral tessellations: {7/3}, {7}s

pentagon out of one, rearrange the three pieces, and cut along three more line segments to give a tessellation element, shown in Figure 16.24. Fourteen of these tessellation elements cover a nonorientable surface formed by gluing together pairs of edges from fourteen {7/3}s. Again, we fold the tessellation element, with the folds corresponding to edges of a {7/3}. The folds are shown as dotted lines in Figure 16.24 and numbered 1 through 6. One face of our nonorientable surface is shown in Figure 16.25, with the edges numbered to correspond to the folds in Figure 16.24.

The dissection is similar to what has just been discussed for the pentagram, except that the {7/3} replaces the {5/2}, and the number fourteen replaces ten. The fourteen {7/3} faces of the nonorientable surface are shown in Figure 16.26. Half of the faces are a rotation of Figure 16.25, and the other half are a reflection of a rotation. We arrange the faces in a symmetric manner, reading the first row from left to right, then the last two in the second row left to right, then the third row left to right, and finally the first two in the second row right to left. We label the pieces in each face with letters from A to N, with each letter labeling all pieces originating from the tessellation element of one of the fourteen copies of the tessellation element.

CHAPTER 17

Farewell, My Lindgren

"Concerning Freese's work, have you tried a letter to the mayor of Los Angeles, saying that from the bits of this work that you have seen, it would really be a great loss to the mathematical world were this outstanding work to be irretrievably lost to posterity? As far as I can see, it could not do any harm, and if you also rub in about the glory of the citizens of his illustrious city, it might produce results."

So is this the MacGuffin of a fast-paced melodrama? The product of a Hollywood script-writer's overactive imagination, destined to launch the career of a mathematical Philip Marlowe, to inaugurate a furious search across three continents, and to lay bare the power politics of a sprawling megalopolitan region?…

[with apologies to Raymond Chandler]

The trail was as cold as the Minnesota tundra that Ernest Irving Freese had fled so many years before. The faded letter from C. Dudley Langford to Harry Lindgren had a 1964 date. It outlined a strategy to obtain the Los Angeles architect's magnum opus on geometric dissections – a strategy as hopeless as announcing that all the sunbaked jurisdictions in Southern California should assemble into one coherent whole. Lindgren had never persuaded Freese's widow to share the manuscript with him. And Freese's blueprints, reportedly containing some 200 plates with over 400 figures, were certainly not on file at City Hall.

Or probably anywhere. Too bad, since the few dissections bootlegged from this now-lost manuscript hold their own under a scorching sun on that rare smog-free day. Freese's work ranged from rational dissections to completing the tessellation to strips to assembling equal figures into larger copies of the same. He was one of the first to use the internal structure of polygons as a basis for dissections. He had fingered a family of polygons whose structure is favorable for finding economical dissections of two equal polygons to one. Was he just a small-time mathematician fronting as an architect? Or did his ideas lead to other families of figures that have *favorable* structure? And to *favorable* dissections beyond the two-to-one examples that I first tracked down?

I jogged my memory and recalled the 4-piece dissection of two equal squares to one in Figure 4.3. It was based on the property that the diagonals of the small squares are equal in length to the side of the large square. Langford (1967a) claimed that Freese was the first to realize that whenever the number of sides of a polygon is a multiple of 4, the internal structure of the polygon contains squares, and thus their diagonals give the side lengths of the larger polygon. A simple method gives a $4n$-piece dissection of two $\{4n\}$s to one. Even a coroner far gone with formaldehyde could cut open octagons by this method (Figure 17.1). Just in case, I sketched the dotted lines to indicate the internal structure. The only cuts not on the boundaries of rhombuses in the internal structure were those on the diagonals of the squares.

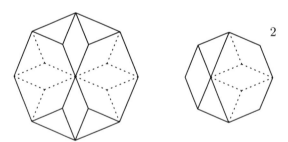

17.1: Freese's two octagons to one

The sun had set, but the heat lingered on. I looked up at the evening sky, thinking of the stars $\{p/q\}$ where p is a multiple of 4. The internal structure of the $\{8/2\}$ contains squares, so that a 12-piece dissection of two $\{8/2\}$s didn't get the drop on me. But it was dead on arrival once Lindgren (1964b) found the elegant 11-piece dissection in Figure 17.2. Again I doodled in the dotted lines to fill out the internal structure.

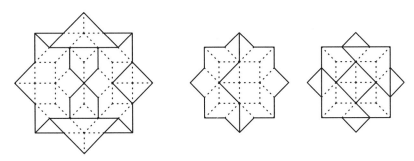

17.2: Lindgren's two {8/2}s to one

Since there are no squares in the internal structure of an {8/3}-star, it shaped up to be a tougher nut. Yet Lindgren (1964b) found the 12-piece dissection of two {8/3}s to one in Figure 17.3. The clue that tipped him off was the isosceles right triangles inside the {8/3}s. The hypotenuse of such a right triangle is $\sqrt{2}$ times the length of a leg.

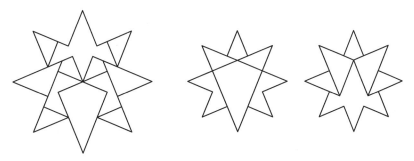

17.3: Lindgren's two {8/3}s to one

I was taking the shortcut over to 221 Assembly Boulevard when the pieces fell into place: Four pieces for squares, and eight pieces for octagons, meant twelve pieces for dodecagons. That was Freese's modus operandi, according to Langford (1967a). Letting down my guard for a moment, I was blind-sided by Lindgren's (1964b) 10-piece dodecagon dissection (Mugshot 17.4). It had horned in on a 13-piece dissection by Joseph Rosenbaum (1947). The damage to Lindgren's dodecagon was relatively minor, because the 60°-rhombus in its structure splits easily into two equilateral triangles. To cover my jitters until the mathic cop showed up, I drew more dotted lines for its internal structure.

The cop was okay, as cops go. He liked things on the square, so I showed him my two-to-one dissections of 12-pointed stars. First, the 13-piece dissection of {12/2}s in Figure 17.5. "Yeh, there's squares in it," he grunted, "but you got a prior in (Frederickson 1972a)."

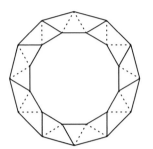
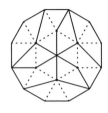

17.4: Two dodecagons to one

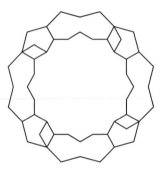
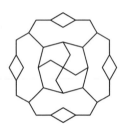

17.5: Two {12/2}s to one

Getting nervous, I spilled the details on Lindgren's (1964b) 16-piece dissection of two {12/3}s to one. I explained that Lindgren had viewed each {12/3} as similar to a square and attempted a dissection related to two squares to one. Then I sprang my 12-piece dissection (Figure 17.6) on him, glossing over my prior in (1972a). The cop eyed me warily when I said that it is based on the same idea. I cut notches in the small {12/3}s so that their halves fit neatly together. Then I cut "halves" to form as much of the outline of the large {12/3} as possible, and I also cut a notch into which I fit the pieces remaining from cutting the other notches. The piece that corresponds to a "half" looks like a geometrical gerrymander. I called it a "geomander." "You private math-dicks is always cracking wise," he growled.

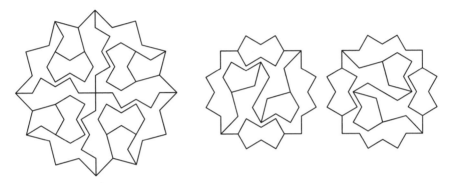

17.6: "Jumpin' Geomanders!": two {12/3}s to one

Quickly I pulled out my {12/4}s – my 18-piece model (1974). He couldn't find the squares in Figure 17.7. I explained – a bit too glibly – that attaching an equilateral triangle to a 30°-rhombus produces a figure with a right angle, from which we can cut an isosceles right triangle. Unfortunately, this cruiser was equipped with cop brains: "Don't see no squares," he frowned. I was lucky to get off with a warning.

17.7: Two {12/4}s to one

191

It was dark now, and I noticed a strange group in tan robes congregating in front of an odd geometric building. With euphoric expressions on their faces, they lined up in threes and entered through a triangular door under a sign that read "The Triangulary." A stone tablet declared, "When the apex is $\frac{1}{2}\sqrt{3}$ above where you have begun, then you will form three into one." Etched beneath the words was the 6-piece dissection of three triangles to one! (Figure 5.7.) The robes had by now all disappeared through the door, and after three moments' hesitation, so did I.

I entered a dimly lit anteroom, beyond which was a main hall where I could see the congregation in prayer. On one wall of the anteroom was a dissection of three stone hexagrams to one. Of course, a hexagram consists only of equilateral triangles! It was the 13-piece dissection by Lindgren (1964b), which he had soon reduced to twelve pieces, in (Lindgren 1964a). Whoever these "robes" were, they hadn't seen my 12-piece dissection (1972a), with 6-fold rotational symmetry in the large star (Figure 17.8).

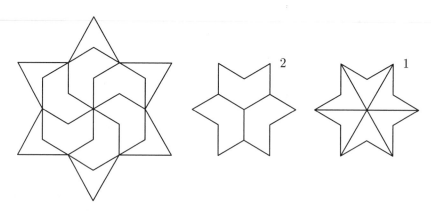

17.8: "Star Bright": three {6/2}s to one

Farther along the wall were large and small enneagons, displaying their internal structure (Figure 17.9). Here the designer had at least discovered an 18-piece dissection, better than the 21-piece dissection in (Cundy and Langford 1960). The dissection glorified the fact that the height of a triangle in the small enneagon gives half the side of the large enneagon.

But these "triangulareans" had committed their marvels to stone too quickly, unaware that I had reduced the number of pieces to 15. I had "borrowed" a triangle adjacent to another triangle, so that a line segment between nonadjacent apexes of these triangles gave the side length for the large enneagon (1972c). And pulling the ultimate squeeze in this borrowing racket, Robert Reid and Anton Hanegraaf had then independently found 14-piece dissections (Figure 17.10).

192

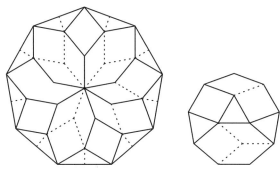

17.9: Enneagon structure

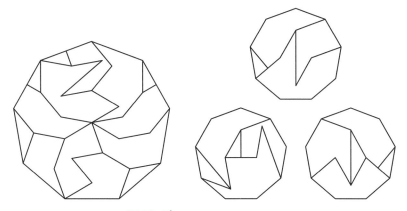

17.10: Three enneagons to one

The next panel contained a 15-piece dissection of three dodecagons to one, like the one in (Lindgren 1964b). Lindgren had left one of the small dodecagons uncut and achieved 2-fold rotational symmetry in the large dodecagon. But again, the triangulareans were ignorant of the independent discoveries by Robert Reid and Anton Hanegraaf, reducing the number of pieces to 14 (Figure 17.11). Farther along the wall I found a small plaque that declared: "Equality, fraternity, and spirituality to he, she, or it, who brings one dodecagon thrice rotated from three similarly split."

Puzzle 17.1: Find a 15-piece dissection of three dodecagons to one such that the large dodecagon has 3-fold rotational symmetry and the small dodecagons have 3-fold replication symmetry.

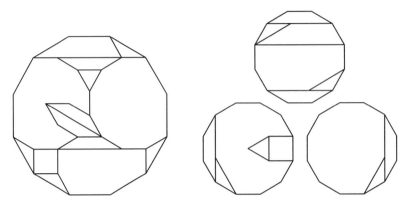

17.11: Variation of Reid's three dodecagons to one

Dissections of 12-pointed stars decorated the opposite wall of the anteroom. The first was of three {12/2}-stars, in Figure 17.12. It was my 18-piece dissection (1972a). Momentarily I grew nostalgic, recalling that I had cut two of the small stars into thirds and then arranged them around most of the third star. I cut the third star into twelve pieces, with six going to fill out the boundary of the large star, and six rotating to fill in the inner cavity of the large star. The large star has a pleasing 6-fold rotational symmetry.

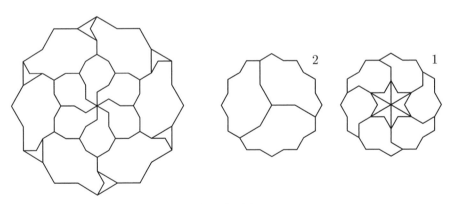

17.12: Three {12/2}s to one

The triangulareans had no dissection for three {12/3}s. Would they have been surprised to learn that I had dissected three {12/3}s to one {12/3} in no more pieces than my three {12/2}s to one? My 18-piece dissection (1974) seemed somehow suggestive of tremendous internal forces that this group could not yet control (Figure 17.13).

17.13: "Internal Cataclysm": three {12/3}s to one

Finally, given a prominent spot on the wall was a 24-piece dissection of three {12/4}s to one. It was identical to my dissection (1972b), and also to Anton Hanegraaf's in (Tjebbes 1969). Overwhelmed by the attractive symmetry, the triangulareans had labeled this one "Supernova." Sadly, they had not realized that Robert Reid had exploded this supernova (Astronomers: Take note!) by designing the remarkable 23-piece dissection in Figure 17.14.

17.14: Reid's three {12/4}s to one

Having completed my inspection of the anteroom, I peeked into the main hall, where a robed leader, no doubt a "robes-Pierre," was working himself into a frenzy. I must have distracted him, for he pivoted 60° toward me and roared, "We have a dissembler in our midst!" The robes turned and glowered at me.

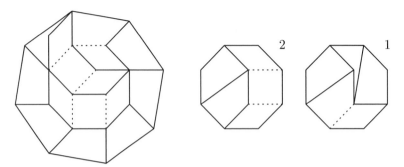

17.15: "Stop, Stop, Stop": three octagons to one

"Stop, stop, stop!" I cut in. "Eight is not a multiple of 3, and yet there is a favorable dissection of three octagons to one! The reason is that $\sqrt{3}$ is the length of a diagonal of a $1 \times \sqrt{2}$ rectangle, and such a rectangle is easily found in an octagon. Just attach isosceles right triangles to the edges of length $\sqrt{2}$, and you get a hexagon formed from one square and two 45°-rhombuses. Here is my 10-piece dissection (1974)." I used dotted lines to show them the internal structure of the dissection in Figure 17.15.

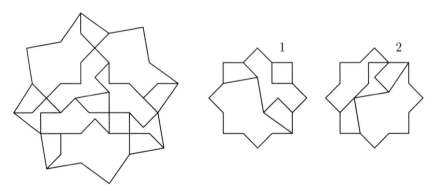

17.16: Reid's three {8/2}s to one

The stunned silence emboldened me: "There is also a favorable three-to-one dissection for an {8/2}! Using the internal structure of the small star, I can place a line segment of length $\sqrt{3}$ times the side length. In fact, I can locate three such line segments in the outline of the large star. I exploited this property to give a 16-piece dissection (1974). Furthermore, my associate Robert Reid has combined this observation with others to create a brilliant 15-piece dissection (Figure 17.16)." Suddenly the floor shook violently and then split. Fire shot up from the central

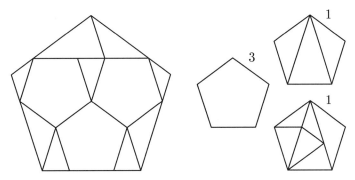

17.17: Langford's five pentagons to one

altar and into the ceiling beams. The building groaned. Robed fright swept me out through the exit.

A lone siren still warbled its warning in the distance. I popped a pair of pento-barbitals into my mouth, lay down in bed, and closed my eyes. Five pentagons danced out behind my eyelids. The ratio of the lengths of a pentagon's diagonal and side is the golden ratio, $\phi = (1 + \sqrt{5})/2$. Since the sum of ϕ and its inverse is $\sqrt{5}$, I should have expected a favorable dissection of five pentagons to one. C. Dudley Langford (1956) had obliged, with his elegant 12-piece dissection (Figure 17.17).

Puzzle 17.2: Find a 15-piece dissection of five pentagons to one, in which the large pentagon has 5-fold rotational symmetry and the small pentagons have replication symmetry and can be hinged.

C. Dudley Langford was born in Highgate, London. He entered the Imperial College in 1923 and received a Ph.D. in mathematics at the age of twenty-two. After a short time in industrial research, he switched to teaching, finding especial satisfaction at the King Alfred "modified Dalton plan" school. Long before "modern math" came into vogue, he had his pupils labor to construct multiplication tables in a number of bases.

Dr. Langford retired after a series of fractures caused by muscular weakness, the effects of polio, which he had first contracted when he was twelve. Over the years, and in spite of his increasing health problems, he was a regular contributor to the *Mathematical Gazette*. A problem he posed shortly before his death in 1969 drew a record response from readers.

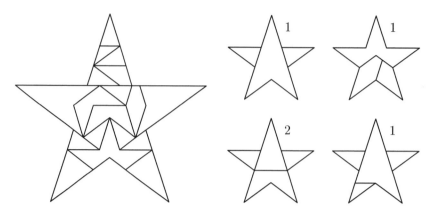

17.18: Five pentagrams to one

The pentagons dissolved, and five pentagrams took their place. Since the same ratios appear in a pentagram as in a pentagon, there should be a simple dissection of five pentagrams to one. Lindgren (1964b) attributed a symmetrical 20-piece dissection to Ernest Irving Freese. But I had done better (1974), finding the 18-piece dissection of Figure 17.18. Was a 17-piece dissection possible?

Like dancers in a Christmas ballet, the pentagrams were brushed aside by five decagons. Langford (1960) had dissected five decagons to one in 25 pieces. The vision of a 20-piece dissection with 5-fold rotational symmetry in the large decagons and 5-fold replication symmetry in the small decagons danced in my head. The 20 pieces reduced easily to 18, but then I seemed to be butting my head against a wall. Finally, a hardheaded 17-piece dissection (Figure 17.19) appeared (1974).

Now deep in sleep, I dreamed of a freezing cop, of five squares in a line-up, and of a cross-town diagonal of length $\sqrt{5}$. Abū'l-Wafā's nifty dissection of $a^2 + b^2$

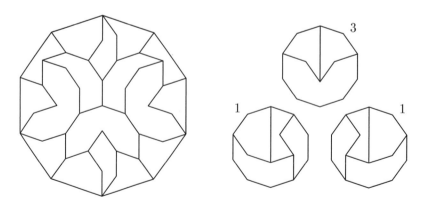

17.19: "Ram's Head": five decagons to one

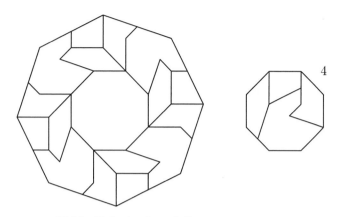

17.20: "Spinning Stops": five octagons to one

squares to one shimmered into view, with $a = 1$ and $b = 2$. I recalled the 9-piece dissection of five squares to one in Figure 6.5. What about favorable five-to-one dissections of polygons and stars that contain squares in their structure? My first was the 17-piece dissection of five octagons to one in Figure 17.20. I pictured a line segment from a corner of a square to the midpoint of a nonadjacent side, of length $\frac{1}{2}\sqrt{5}$ times the side length of the square. For economy, I swapped right triangles whose hypotenuse is this length between adjacent pieces. The large octagon blurred as it spun around in my head.

My dream spun on to five {8/2}s. Small squares popped out to make them fit, and cuts across two squares in a row gave the outer edges. The result was the

17.21: Five {8/2}s to one

199

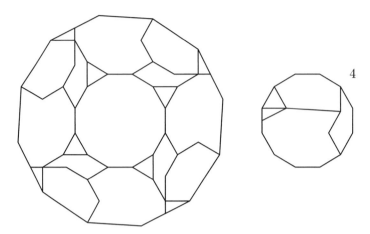

17.22: Five dodecagons to one

21-piece dissection in Figure 17.21. As in the octagon dissection, there were 4-fold rotational symmetry and translation with no rotation.

My fivefold slumbers drifted on to dodecagons. I dreamt of two squares side by side in the dodecagonal interior, their combined diagonal giving the side of a large dodecagon. My 21-piece dissection floated into view (Figure 17.22). It spun around with 4-fold rotational symmetry but would not translate.

The phone had rung six times before my sleep-befuddled hand had somehow found the receiver. A muffled voice had rasped, "The Splitwoods can help you with Freese's...," and the line got cut off. My recollection was also cut off, by a peremptory "The General will see you now in the Dudenerium." General Splitwood, a sharp-eyed sexagenarian, wasted no time with pleasantries: "So this Freese fellow knew

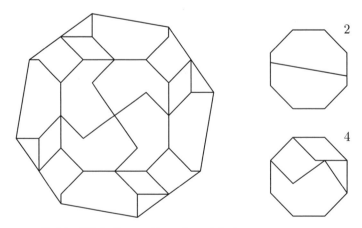

17.23: "High-Octane Houndstooth": six octagons to one

about two to one, and you think you know about three to one. So did you pool your resources to get six to one?" Stopped short, I had to admit that I had the 20-piece dissection of octagons in Figure 17.23. "Translational! And a nice houndstooth," volunteered the General.

Puzzle 17.3: Find a 20-piece dissection of six octagons to one in which four of the small octagons remain uncut.

The butler returned with a plate of fruit for the general, who grimaced, "Six dodecagonal slices – does it matter how you slice them?" I wondered whether he remembered my ungainly 24-piece dissection of six dodecagons to one (1972c). Deftly I showed him my newer (1974) design (Figure 17.24). The old warrior perked up for a moment: "6-fold rot and 6-fold rep." But he faded when I reminded him of Freese, and I had to find my own way out.

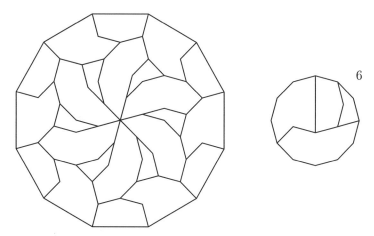

17.24: "Sliced Peach Salad": six dodecagons to one

A pair of fluttering eyelashes stopped me in the hall. "You're cute! And you know twosies and threesies. But can you play 'two-into-three'?" She probably didn't have a license, so instead I cut up with my (1974) dissection of three octagons to two (Figure 17.25).

She persisted, "I'm cute too, and I know all about three dodecagons to two. And how Lindgren (1964b) gave a 23-piece dissection, which you reduced to 22 pieces (1974). And how you got the pants beat off you by Robert Reid when he reduced the number of pieces to 18." Reid had cleverly assembled a small square

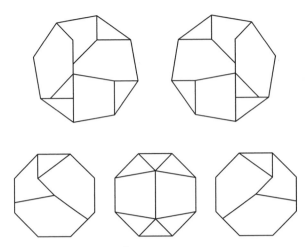

17.25: Three octagons to two

plus a 30°-rhombus into a larger square in five pieces but had turned over two of the pieces. She hadn't seen my variation of it in Figure 17.26, which turns over no pieces. The trick was to assemble the small square and rhombus by using a P-slide. With one cut I transformed the rhombus into a rectangle twice as high as it is wide and positioned it next to the square before making the P-slide. To avoid using an extra piece, I had attached a part of a rhombus to the small square before making the cut in the P-slide and then had filled in the borrowed space.

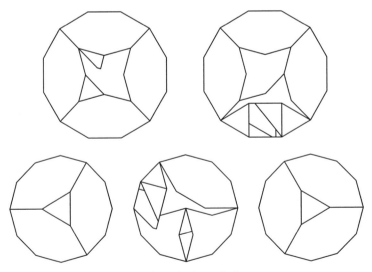

17.26: Variation of Reid's three dodecagons to two

Robert Reid was born in England in 1926. He won a scholarship to Kings School, Canterbury, where he excelled in mathematics, and from which he graduated in 1944. After the Second World War, he went to Peru for two years before enrolling in Queens University in Ontario, Canada. A promising undergraduate career in mathematics was cut short by illness, and he returned to Peru in 1950. Residing there since then, he has managed an import firm; has owned and operated a cinema; has been a supplier of properties for TV, cinema, and theater; and has more recently been an antiques dealer.

A constant throughout Robert's life has been his love of mathematics, especially of the recreational variety. He has been writing two books, one on dissections and another on tessellations. Prolific and audacious, he claims to be planning his own version of the *JRM*, namely, the *Journal of Reid's creations in Mathematics*!

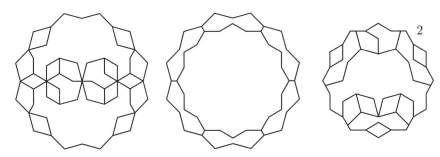

17.27: Elliott's three {12/2}s to two

Spooked, she bounded off, and I was left alone to ruminate on Stuart Elliott's (1983) 27-piece dissection of three {12/2}-stars to two (Figure 17.27).

My last hope was a friend at the Seventh Heaven Café in Venice West. Still waiting for his beard to show up for work, he sat at a table near the window and scribbled in a spiral notebook. Except for him, the place was empty. The crowds would come later. My greeting of "How's the heptameter?" elicited a sardonic smile.

He listened vaguely to a narrative of my progress and then began his own narrative, on dissections of seven figures to one. "I'll lay it on you, man: Like when these triangles make the scene, then you know you're hip, 'cause there'll be a diagonal whose length is $\sqrt{7}$. This Alfred Varsady is one crazy cat. Just take seven and relate to his hexagrams." I admired Alfred Varsady's 19-piece dissection (Figure 17.28).

"Varsady's next dissection is a gas: seven – count them – enneagons to one. He hustles one of the small enneagons into the center of the large one, then has the pieces from the others fall in around it. Can you dig it?" Varsady's approach

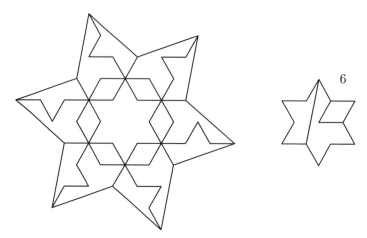

17.28: Varsady's seven hexagrams to one

was inspired, but he had cut out a little too soon. Following an idea similar to that in Figure 17.10, I borrowed a triangle and combined it with another small triangle to get half of the side length of the large enneagon. The result was the 28-piece dissection in Figure 17.29.

"Hey, now you're swinging. But this Varsady, man, is like avant-garde. This next one's gonna flip your wig." He sketched out an amazing dissection of seven heptagons to one. How had Varsady found the $\sqrt{7}$ lurking in a heptagon? The trigonometric relation $\sqrt{7} = \tan(\frac{3}{7}\pi) - 4\sin(\frac{1}{7}\pi)$ implies that there should be a favorable dissection of seven heptagons to one. Varsady positioned a cut of length $\sqrt{7}/2$ times

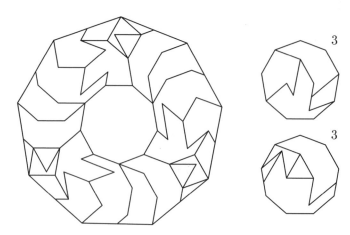

17.29: Seven enneagons to one

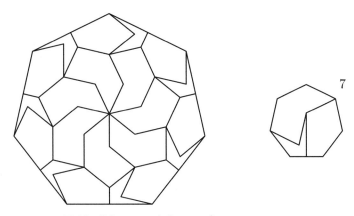

17.30: "Heptomania": seven heptagons to one

the length of a side in the middle of each small heptagon. But again, he had cut out too soon, and I could adapt his approach and reduce the number of pieces from 35 to 21. I identified two pieces on each side of the cut that fold around to bound one side of the large heptagon, and such that just one piece remains. The remaining piece then forms one-seventh of the starlike area in the interior of the large heptagon (Figure 17.30). I caught the guy stealing a second, a third,..., a seventh look: "Now that is hep.... Hept-fold."

"Okay, so pick up on this one. If seven heptagons send you, then how about eight? Alfred Varsady is ... far out." I compared Varsady's new dissection with his last one. The cut of length $\sqrt{7}/2$ bisects one of the sides of the heptagon. Thus the line segment from the top of the cut to an endpoint of that side is of length $\sqrt{2}$.

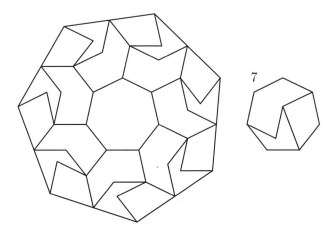

17.31: Eight heptagons to one

205

Since $\sqrt{2}$ is half of $\sqrt{8}$, Varsady had realized that this leads to a favorable dissection of eight heptagons to one. Then I modified his 29-piece dissection to give the 22-piece dissection in Figure 17.31. I cut seven of the heptagons in a fashion similar to that in Figure 17.30 and left the eighth uncut. There was 7-fold rotational symmetry as in the previous figure, but now seven pieces are turned over.

"Man, that beats 77 strips driving off in the sunset. So I bid thee a fond farewell, my Lindgren." I wasn't sure I had caught the allusion but didn't try too hard, since an offbeat thought had distracted me. The appearance of the $\sqrt{2}$ in the heptagon also implies that there should be a favorable dissection of two heptagons to one. I got in the groove and produced the 10-piece dissection in Figure 17.32. I cut the heptagon on the right as the seven in the previous figure, and I started the middle heptagon that way. But I had to fill a triangular space next to the bottom edge in the large heptagon, using an isosceles triangle from the top of the middle heptagon. I used a T-strip dissection to convert one triangle to the other and saved a piece by swapping a triangle with the piece below to accommodate the T-strip pieces better. This cost one piece but saved two.

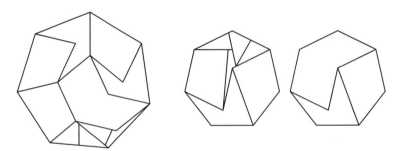

17.32: Two heptagons to one

Now I was hot. I had made the scene with three triangles to one, five pentagons to one, and seven heptagons to one. Nine enneagons to one was no sweat, since $\sqrt{9} = 3$. What about eleven {11}s to one? I recalled that $\sqrt{3} = \tan(\frac{1}{3}\pi)$. After comparing with the preceding equation for $\sqrt{7}$, I found $\sqrt{11} = \tan(\frac{5}{11}\pi) - 4\sin(\frac{4}{11}\pi)$ by trial and error. Yet this did not seem to lead to any dissection as natural and as pretty as that in Figure 17.30. As in the heptagon, I could bootstrap the appearance of the $\sqrt{11}$ in an {11} into appearances of $\sqrt{3}$ and $\sqrt{12}$. However, my friend found this a drag and weighed in with the last word:

"You're a hep cat. But that two-to-one jazz is for squares. And if you're turning on with {11}s, you'd better split before you get us both busted. Can't help you with the Freese, man.... But that's cool."

CHAPTER 18

The New Breed

Three fat folders lie on the desk. Within each is a series of enthusiastic letters accompanied by a stack of remarkable diagrams. First is the correspondence from a now-deceased Englishman, second is that from an expatriate Hungarian, and the third is from an expatriate Englishman. Together, these dog-eared and annotated sheets document the origins of a new type of dissection. It has been a wonderful experience to receive these letters, overflowing with new dissections, new ideas, written by new people. They tell the story of expanded boundaries, not merely of countries, not only of dissections, but also of human possibilities. A memorable record, in three fat folders.

The folders are for Stuart Elliott, Alfred Varsady, and Robert Reid; the first two reach back to the late 1970s, and the last commences with a decade-old letter to Martin Gardner. These manila cornucopias chronicle the development of new types of dissections by people who passionately mapped the new terrain. Elliott led off with dissections based on combinations of the relationships discussed in the last two chapters. Soon after, Varsady developed a system of relationships based on decompositions of figures to half-rhombuses. Finally, Reid was drawn into dissections as he whittled down dramatically the number of pieces in many of Elliott's dissections. These three demonstrated that Harry Lindgren's work on the internal structure of polygons and stars could be pushed in exciting new directions.

So we begin with Stuart Elliott. In Chapter 16, we saw special relationships that led to auspicious dissections among certains pairs of stars and polygons. In Chapter 17, we explored favorable dissections of n identical polygons to one. Elliott realized that the two types of relationships could be exploited together to give a "new breed": what we call *propitious* dissections.

As we saw in Figure 10.21, relationship (16.1) is the basis for an auspicious dissection of a {6/2} to a {3}. Since there is also a favorable dissection of three {6/2}s to one {6/2}, Elliott (1983) looked for a propitious dissection of three {6/2}s to one {3}. He found a simple 9-piece dissection that has 3-fold replication symmetry in the hexagrams and 3-fold rotational symmetry in the triangle. But never satisfied with something simple and symmetrical, Reid (1987) discovered a clever 8-piece dissection (Figure 18.1). In this case, the combined relationship is so simple that all edge lengths are integral multiples of one given length.

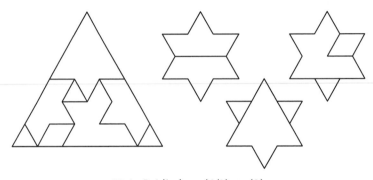

18.1: Reid's three {6/2}s to {3}

As another example, we have seen a favorable dissection of two {12/2}s to one in Figure 17.5, and an auspicious dissection of a {12/2} to a {6/2} in Figure 16.11. Thus Elliott (1983) sought a propitious dissection of a {12/2} to two {6/2}s. His 13-piece dissection appears in Figure 18.2.

Similarly, we have seen an auspicious dissection of a {12/2} to a {6}, and a favorable dissection of two {12/2}s to one. Again Elliott (1983) found a propitious

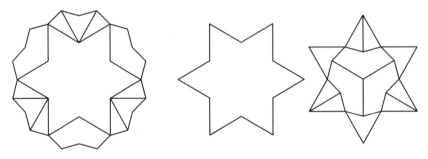

18.2: Variation of Elliott's {12/2} to two {6/2}s

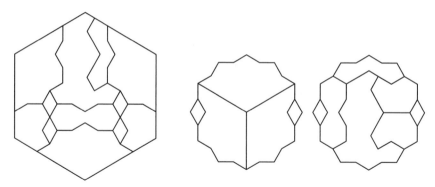

18.3: Variation of Reid's two {12/2}s to {6}

dissection of two {12/2}s to one {6}. This appeared to be one of the few of Elliott's dissections that Reid couldn't touch. But with some clever swapping, Robert produced a surprise that saved a piece. We give a minor rearrangement of his 12-piece dissection in Figure 18.3.

We can use more than two relationships to derive propitious dissections. For example, Elliott (1983) combined the two relationships in the previous example with three {6}s to one to give two {12/2}s to three {6}s. He gave a beautifully symmetric 18-piece dissection, which prompted Robert Reid to find a quite different dissection in seventeen pieces. But this time I surprised Robert with the 15-piece dissection in Figure 18.4. We can generate this dissection by superposing tessellations. The trick is to find a tessellation for the {12/2}s that works well with a tessellation of hexagons.

Puzzle 18.1: Draw the tessellation element for the dissection of two {12/2}s to three {6}s.

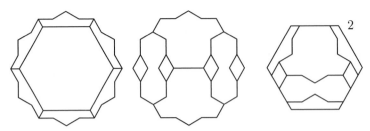

18.4: Two {12/2}s to three {6}s

Another beautiful example in which Reid bettered Elliott is the dissection of three {6/2}s to one {12/2}. It results from combining the relationship for a {12/2} to a {6/2} together with the one for three {6/2}s to one. Elliott (1983) gave a 15-piece dissection, but Reid found a clever 13-piece dissection (Figure 18.5).

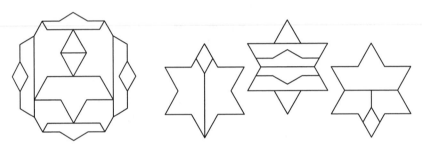

18.5: Variation of Reid's three {6/2}s to {12/2}

Elliott also found that he could use three relationships involving three different figures, where one of the figures does not appear in the final dissection. He combined the relationships for a {12/2} to a {6}, three {12/2}s to one, and two {6/2}s to one {6}, to suggest a propitious dissection of three {12/2}s to two {6/2}s. He gave a 36-piece dissection, which Reid (1987) reduced dramatically to 27 pieces. Not to be left out, I found a symmetric 24-piece dissection, but Robert Reid weighed in again with a 23-piece dissection, a variation of which is in Figure 18.6. Two of the {12/2}s are cut into eight pieces, including cuts for the dashed edges, and the third is cut into seven pieces, ignoring the dashed edges.

We saw a natural dissection of a dodecagon to three squares in Chapter 1. This dissection is actually propitious, based on a favorable dissection of three dodecagons to one, and an auspicious dissection of a dodecagon to a square. Since a {12/5} is readily transformed to two dodecagons, and a Latin Cross consists of six squares, there should be a propitious dissection of a {12/5} to a Latin Cross.

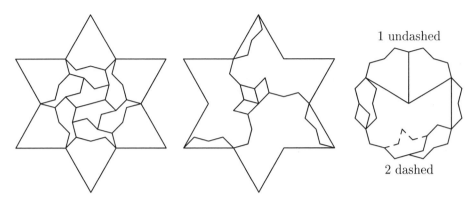

1 undashed

2 dashed

18.6: Reid's three {12/2}s to two {6/2}s

Lindgren (1964b) produced his stunning 7-piece dissection, which has already daz-
zled us (Figure 3.1).

The preceding examples barely scratch the surface of economical dissections
based on Elliott's principle. Reid has improved on many of Elliott's dissections
and produced many additional dissections of the same type. Encouraged by their
success, we identify a new propitious dissection that has been a challenge.

We began this book with the auspicious dissection of a 7-pointed star to a pair
of heptagons. Could we hope for a propitious dissection of that same 7-pointed
star to a single heptagon? Since we found two heptagons to one to be favorable
in the last chapter, we can combine this with the {7/3} to two heptagons to get
what we want. My 14-piece dissection of a {7/3} to a {7} appears in Figure 18.7. The
triangle cut from the left middle point in the star is not an isosceles triangle. Just
as in the dissection of two heptagons to one, we convert an isosceles triangle in

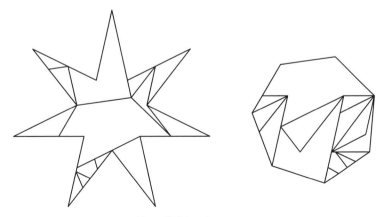

18.7: {7/3} to heptagon

211

the lower left point to an irregular triangle on the lower right of the heptagon. But this time we do not avoid using a full four pieces. Can a reader find a dissection with fewer pieces?

Exploring in a different direction, Alfred Varsady (1986) studied how to build 5-pointed and 10-pointed figures out of two basic building blocks shown in Figure 18.8. The first is $\{3_{1/5}\}$, an isosceles triangle with two angles equal to $\pi/5$, and the second is $\{3_{2/5}\}$, an isosceles triangle with two angles equal to $2\pi/5$. I call these the *iso-penta triangles* and keep a count of their number in any figure as a pair: For example, as shown in Figure 2.2, a pentagon contains $(3, 1)$, that is, three $\{3_{1/5}\}$s and one $\{3_{2/5}\}$. From Figure 2.6, a pentagram contains $(2, 4)$: two $\{3_{1/5}\}$s and four $\{3_{2/5}\}$s. A decagon contains $(10, 10)$. Two pentagons and two pentagrams supply the right number of iso-penta triangles for a decagon, and there is a dissection to match, as shown in Figure 18.10. We term such a dissection *advantageous*.

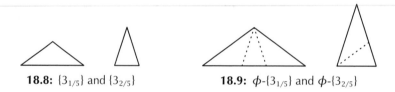

18.8: $\{3_{1/5}\}$ and $\{3_{2/5}\}$ **18.9:** ϕ-$\{3_{1/5}\}$ and ϕ-$\{3_{2/5}\}$

Varsady then used a simple expansion to give versions of $\{3_{1/5}\}$ and $\{3_{2/5}\}$ that have sides longer by a factor of ϕ (the golden ratio). Figure 18.9 shows that two $\{3_{1/5}\}$s plus one $\{3_{2/5}\}$ form a larger $\{3_{1/5}\}$, and one each of $\{3_{1/5}\}$ and $\{3_{2/5}\}$ similarly form a larger version of $\{3_{2/5}\}$. Thus we can form ever larger versions of $\{3_{1/5}\}$ and $\{3_{2/5}\}$.

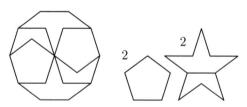

18.10: Two $\{5\}$s plus two $\{5/2\}$s to $\{10\}$

Applying this principle, Varsady produced some wonderful decagon dissections. A ϕ-decagon consists of $(10, 10)$, which when expanded gives $(30, 20)$. These iso-penta triangles constitute eight pentagons and three pentagrams. Varsady's corresponding dissection is given in Figure 18.11.

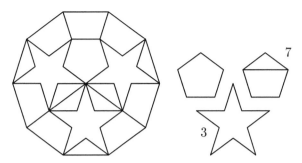

18.11: Varsady's eight {5}s plus three {5/2}s to {10}

Not stopping there, Varsady recognized that he could perform the substitutions yet again. A ϕ^2-decagon has $(80, 50)$, or equivalently 22 pentagons and 7 pentagrams. Varsady's corresponding dissection is given in Figure 18.12. We can show that the number of pentagons at one level is twice the number of pentagons and pentagrams at the previous level. Similarly, the number of pentagrams is equal to the number of pentagrams plus half the number of pentagons at the previous level. These three dissections are members of a family, with the next member being a ϕ^3-decagon, composed of $(210, 130)$, or equivalently 58 pentagons and 18 pentagrams.

Varsady constructed another natural family in the following way: He decomposed a $(\sqrt{\phi}/\sqrt[4]{5})$-decagon into two pentagons and one pentagram, which together have $(8, 6)$. His corresponding dissection (1986) is shown in Figure 18.13. Expanding

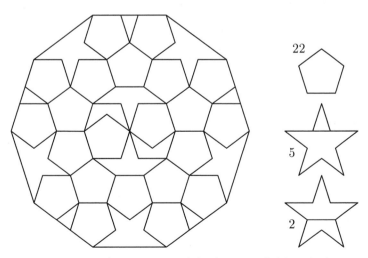

18.12: Varsady's twenty-two {5}s plus seven {5/2}s to {10}

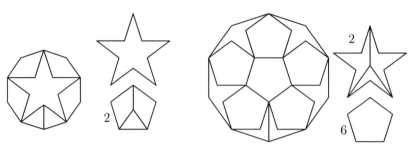

18.13: Two {5}s, one {5/2}   **18.14:** Six {5}s, two {5/2}s

each dimension by a factor of ϕ gives $(22, 14)$. A dissection of the corresponding six pentagons and two pentagrams to a $(\phi^{1.5}/\sqrt[4]{5})$-decagon is shown in Figure 18.14. This dissection was independently discovered by Robert Reid.

One more expansion gives $(58, 36)$. A dissection of the corresponding sixteen pentagons and five pentagrams to a $(\phi^{2.5}/\sqrt[4]{5})$-decagon is shown in Figure 18.15.

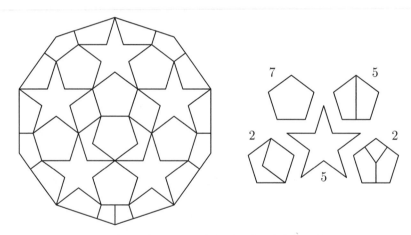

18.15: Varsady's sixteen {5}s plus five {5/2}s to {10}

These decagon dissections are just the tip of the iceberg: Alfred Varsady has gone on to produce a wide range of advantageous dissections. In any of the relationships he discovered, he could replace a decagon with a pentagram, or vice versa. In fact, the second of the two decagon families seems more natural for pentagrams, since a ϕ-pentagram has $(8, 6)$, a ϕ^2-pentagram has $(22, 14)$, and so on. Another example is the substitution of a {10/3} for two {5/2}s. From Figure 2.5, it is easy to see that a {10/3} has $(20, 20)$. Thus a ϕ-{10/3} has $(60, 40)$. Since a {5/2} has $(2, 4)$, a ϕ^3-{5/2} has $(58, 36)$. It follows that a ϕ^3-{5/2} plus a {5/2} give a ϕ-{10/3}. Varsady found a 12-piece dissection, which Reid modified to 11 pieces in Figure 18.16.

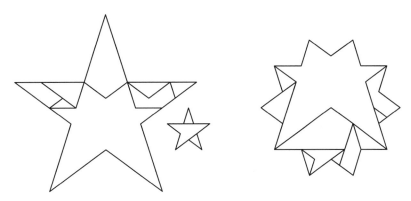

18.16: Reid's ϕ^3-{5/2} plus {5/2} to ϕ-{10/3}

A pentagon has $(3, 1)$, a ϕ-pentagon has $(7, 4)$, a ϕ^2-pentagon has $(18, 11)$, and a (2ϕ)-pentagon has $(28, 16)$. Taking advantage of these relations, Varsady (1986, 1989) gave 7-piece dissections of pentagons realizing $1^2 + \phi^2 + (\phi^2)^2 = (2\phi)^2$. One of these is shown in Figure 18.17.

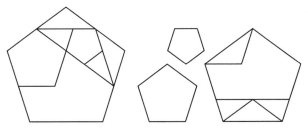

18.17: Varsady's pentagons for $1^2 + \phi^2 + (\phi^2)^2 = (2\phi)^2$

Similarly, a pentagram has $(2, 4)$, a ϕ-pentagram has $(8, 6)$, a ϕ^2-pentagram has $(22, 14)$, and a (2ϕ)-pentagram has $(32, 24)$. Alfred Varsady had found a 13-piece dissection for pentagrams, realizing $1^2 + \phi^2 + (\phi^2)^2 = (2\phi)^2$. This was improved by Robert Reid in Figure 18.18.

Varsady also discovered that more complex combinations of basic triangles are possible. A $(\phi\sqrt{5})$-{$3_{1/5}$} has $(10, 5)$, and a $(\phi\sqrt{5})$-{$3_{2/5}$} has $(5, 5)$, as shown in Figure 18.19. Trivially, four of each triangle make a larger copy, which is also (hidden) in the same figure: the top four small triangles in the large triangle on the left, and the rightmost four small triangles in the large triangle on the right. Thus a $(2\phi^3)$-pentagram has $4 * (58, 36) = (232, 144)$, and a $(\phi^3\sqrt{5})$-pentagon has $5 * (47, 29) = (235, 145)$. A 1-pentagon provides the difference of $(3, 1)$. Varsady's 8-piece dissection, in Figure 18.20, is a marvelous demonstration of this identity.

18.18: Reid's pentagrams for $1^2 + \phi^2 + (\phi^2)^2 = (2\phi)^2$

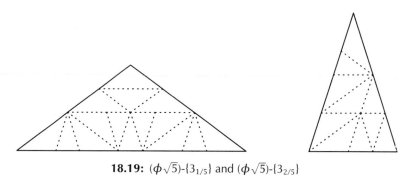

18.19: $(\phi\sqrt{5})$-$\{3_{1/5}\}$ and $(\phi\sqrt{5})$-$\{3_{2/5}\}$

There are nifty, unexpected relationships among pentagons that Varsady has uncovered. We know that a ϕ^2-pentagon has $(18, 11)$. From Figure 18.19, we deduce that a $(\phi\sqrt{5})$-pentagon has $(35, 20)$. A $(\sqrt{4-\phi})$-pentagon, which has $(8, 1)$, results from taking a pentagon, arranging five $\{3_{1/5}\}$s around it, and dissecting them, as

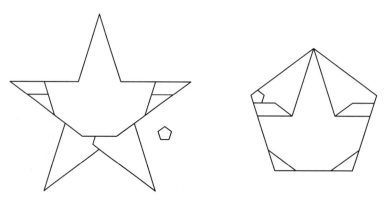

18.20: Varsady's $(2\phi^3)$-$\{5/2\}$ plus $\{5\}$ to $(\phi^3\sqrt{5})$-$\{5\}$

shown in Figure 6.18. Thus a $(\phi\sqrt{4-\phi})$-pentagon contains $(17, 9)$. Dividing through by ϕ gives the relationship $(\sqrt{4-\phi})^2 + \phi^2 = (\sqrt{5})^2$, for which Varsady gives a 7-piece dissection. The dotted lines in Figure 18.21 indicate some of the underlying structure of which he takes advantage.

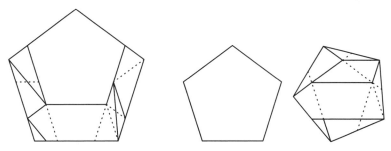

18.21: Varsady's pentagons for $(\sqrt{4-\phi})^2 + \phi^2 = (\sqrt{5})^2$

An unparalleled gem is Varsady's beautiful 11-piece dissection of two pentagrams to one (Figure 18.22). We can view his dissection as the trigonometric restatement of the Pythagorean theorem: $(\sin\theta)^2 + (\cos\theta)^2 = 1^2$. The corresponding right-angled triangle is the left portion of the top point in the largest star, with $\theta = \frac{2\pi}{5}$ the angle on the left in that triangle.

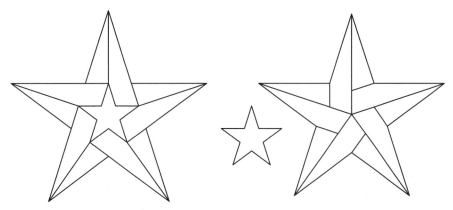

18.22: Varsady's pentagrams for $(\sin\frac{2\pi}{5})^2 + (\cos\frac{2\pi}{5})^2 = 1^2$

Focused primarily on 5- and 10-pointed figures, Varsady missed a neat pattern. A surprising consequence of this last characterization is that similar, economical dissections of other stars are possible. For stars of the form $\{p/2\}$, there are dissections using $2p+1$ pieces. When $p = 6$ and $p = 8$, there are more economical

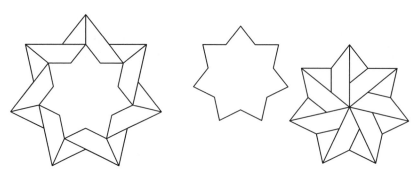

18.23: Heptagrams for $(\sin\frac{2\pi}{7})^2 + (\cos\frac{2\pi}{7})^2 = 1^2$

dissections: See Figures 5.25 and 17.2. For $p = 7$, I have not found a better solution than the 15-piece one of Figure 18.23. An attractive feature is that we can make variously, partially hinged models for the dissections in Figures 18.22 and 18.23.

The structural approach that works for 5- and 10-pointed figures also works for 7- and 14-pointed figures. First, let's find a magic ratio. Let $\vartheta = 2\cos(\pi/7) \approx 1.802$. In a heptagon, the ratio of the length of the shorter of the two diagonals to the length of a side equals ϑ, and the ratio of the length of the longer of the two diagonals to the length of a side equals $\vartheta^2 - 1$. In analogy to the golden ratio in pentagons, I call ϑ the *silvern ratio*; ϑ is also a solution to $y^3 - y^2 - 2y + 1 = 0$.

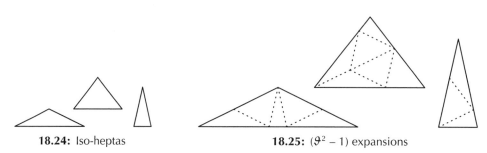

18.24: Iso-heptas   **18.25:** $(\vartheta^2 - 1)$ expansions

For 7- and 14-pointed figures there are three basic building blocks, shown in Figure 18.24. The first is $\{3_{1/7}\}$, an isosceles triangle with two angles equal to $\pi/7$; the second is $\{3_{2/7}\}$, an isosceles triangle with two angles equal to $2\pi/7$; and the third is $\{3_{3/7}\}$, an isosceles triangle with two angles equal to $3\pi/7$. We call these the *iso-hepta triangles*. Figure 2.2 shows how to construct a heptagon from five $\{3_{1/7}\}$s, three $\{3_{2/7}\}$s, and one $\{3_{3/7}\}$. For simplicity, we describe this as the triple $(5, 3, 1)$.

In a manner similar to that for the iso-penta triangles, we can replicate the iso-hepta triangles in larger versions. (ϑ^2-1)-$\{3_{1/7}\}$, (ϑ^2-1)-$\{3_{2/7}\}$, and (ϑ^2-1)-$\{3_{3/7}\}$ are shown in Figure 18.25; $(\vartheta+1)$-$\{3_{1/7}\}$, $(\vartheta+1)$-$\{3_{2/7}\}$, and $(\vartheta+1)$-$\{3_{3/7}\}$ are shown

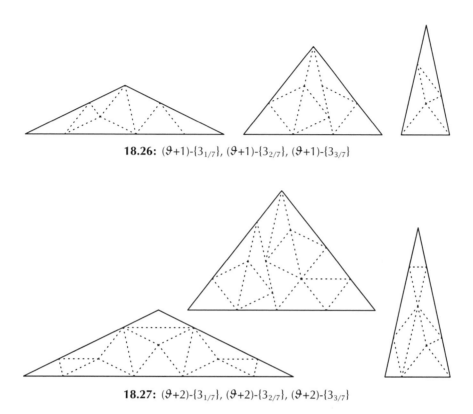

18.26: $(\vartheta+1)$-$\{3_{1/7}\}$, $(\vartheta+1)$-$\{3_{2/7}\}$, $(\vartheta+1)$-$\{3_{3/7}\}$

18.27: $(\vartheta+2)$-$\{3_{1/7}\}$, $(\vartheta+2)$-$\{3_{2/7}\}$, $(\vartheta+2)$-$\{3_{3/7}\}$

in Figure 18.26; and $(\vartheta+2)$-$\{3_{1/7}\}$, $(\vartheta+2)$-$\{3_{2/7}\}$, and $(\vartheta+2)$-$\{3_{3/7}\}$ are shown in Figure 18.27. Alfred Varsady gave a drawing of $(\vartheta+2)$-$\{3_{1/7}\}$ in this last figure, which suggested the likelihood of the other expansions. I have not yet searched for applications of these expansions to identify dissections of the same type as small pentagons plus small pentagrams to a large decagon.

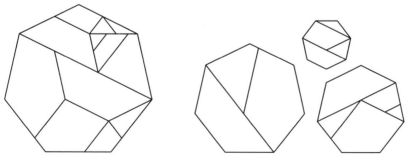

18.28: Varsady's heptagons for $1^2 + \vartheta^2 + (\vartheta^2-1)^2 = (\vartheta^2 + \vartheta - 2)^2$

However, these expansions imply a relationship among heptagons. Varsady has discovered a remarkable 12-piece dissection of heptagons realizing $1^2 + \vartheta^2 + (\vartheta^2 - 1)^2 = (\vartheta^2 + \vartheta - 2)^2$, shown in Figure 18.28. At first glance, it appears that the previous expansions do not help. But the key is to use more than one application of the expansions. Generate a $(\vartheta + 1)$-heptagon, a $(\vartheta^2 - 1)(\vartheta^2 - 1)$-heptagon, a $(\vartheta^2 - 1)(\vartheta + 1)$-heptagon, and a $(\vartheta^2 - 1)(\vartheta + 2)$-heptagon. Dividing the side length of each of the four heptagons by $(\vartheta + 1)$ gives the dimensions as stated in Figure 18.28. Forming the count of triangles for each of the four (expanded) heptagons gives $(26, 31, 15)$, $(83, 103, 46)$, $(129, 160, 72)$, and $(238, 294, 133)$. Sure enough, the sum of the first three vectors equals the fourth.

Alfred Varsady was born in Szeged, Hungary, in 1920. As a school child of twelve, he saw the 4-piece dissection of a triangle and square that had been published by Dudeney. He completed his schooling in 1939, and was assigned to the cartographic institute in Budapest as a map illustrator during the Second World War. After the war, he relocated in Germany in 1945. Since then Alfred has been a technical draftsman and a technical designer, living in Metten, a small village not far from the Czech Republic. Over the past two decades, he has done things with pentagons and pentagrams that would have put a twinkle in Kepler's eye.

CHAPTER 19

When Polygons Aren't Regular

In the early nineteenth century, two lingering difficulties in Euclid's Elements *were addressed and resolved. The first was the troubling role of the parallel postulate. Three men, from Germany, Hungary, and Russia, worked in parallel, effectively decomposing and then recomposing geometry. They showed that the postulate is just one of several valid assumptions that can be used in deductive geometry. This at first caused widespread consternation, which metamorphosed over time into acclamation.*

The second difficulty was to prove the converse of Euclid's assertion that two polygons have equal area if it is possible to decompose one into pieces that recompose to form the other. Three different men, from Germany, Hungary, and England, worked in parallel to decompose and then recompose polygons. But their result caused no consternation and produced at best fleeting fame. Too bad for dissections that the parallel was not stronger!

The non-Euclidean hyperbolic geometry of Carl Friedrich Gauss, János Bolyai, and Nikolay Lobachevsky first shocked and then revolutionized the mathematical world. The same cannot be said of the dissection work of P. Gerwien, Farkas Bolyai, and William Wallace. But their work does give us a method of dissecting any irregular polygon to any other irregular polygon of equal area, using a finite number of pieces. The methods of Lowry (1814), Wallace (1831), Bolyai (1832), and Gerwien (1833) always work, though they often produce a great abundance of pieces.

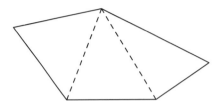

19.1: Partition to triangles

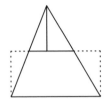

19.2: Triangle to rectangle

Lowry (1814) provided a simple explanation in answer to a problem posed by William Wallace around 1808. (Wallace presumably had a solution at that time, which he gave in expanded form in (1831).) Lowry first divided one of the polygons into triangles, using its diagonals (Figure 19.1). He then converted each triangle to a rectangle, using the construction in Figure 19.2. Next he converted each rectangle to a rectangle of a given length, say ℓ, and stacked the rectangles of length ℓ. Finally, he performed the same set of operations for the other polygon and overlaid the two stacks, which will be the same height. He projected all cuts from one stack to the other, so that the resulting set of pieces will form either of the original two figures.

The conversion of one rectangle R to a longer rectangle R' followed the method of Montucla, as described in (Ozanam 1778). Although we could view it as a P-strip dissection, we describe it here in simpler terms. Extend the base of R to the right and place lines perpendicular to it at intervals equal to the length of the base of R. Draw R' so that its upper left corner coincides with the upper left corner of R and its upper right corner falls on the extension of the base of R. Place a cut in R' wherever a line crosses it. An example is shown in Figure 19.3, where the packing of pieces in R is completed with dotted lines.

Having described a completely general, but not particularly economical, dissection method, we focus in the rest of this chapter on irregular polygons for which we can find more economical dissections.

M. J. Cohn (1975) posed the problem of dissecting a triangle of any shape to a square. For convenience, he assumed that the triangle had area equal to 1 and

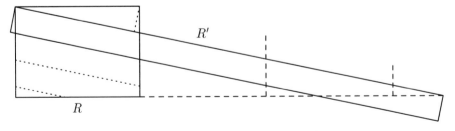

19.3: Montucla's rectangle to rectangle

long side equal to z. He gave a method that uses at most $10 + \sqrt{z^2 - 1}$ pieces and proved two lower bounds on the number of pieces. The first one, of $z/\sqrt{2}$, is easy to see: No piece can have a side longer than $\sqrt{2}$, the length of the diagonal of the square. Thus there must be at least $z/\sqrt{2} \approx 0.7071z$ pieces along the long side of the triangle. By a more complicated argument, Cohn proved a better lower bound, of approximately $0.7415z$. He conjectured, but could not prove, that the number of pieces needed is approximately z.

There are several simpler dissection techniques that use fewer than $5 + z$ pieces. An elementary but cute discovery of mine is that a combination of two techniques does a bit better, using fewer than $4 + z$ pieces. The first technique is best when z is an integer or $\lfloor z \rfloor$ is even (where $\lfloor z \rfloor$ denotes the largest integer no greater than z): Bisect the two shorter sides with cuts perpendicular to the longer side

---

Jean Etienne Montucla was born in Lyons, France, in 1725. Educated at the Jesuit's College in Lyons, he afterward studied law in Toulouse and in Paris. In 1758 Montucla published his book *Histoire des Mathématiques*, which contained not only historical material but also essays on major topics in science. He then served in government, sent first to Grenoble as intendant secretary, then on the expedition for colonizing Cayenne as secretary and royal astronomer. After several years he returned to France as "premier commis des bâtiments"; he also received the honorary appointment as royal censor of mathematical works. Montucla edited the book by Ozanam (1778), adding much new material. His role as editor was so well concealed that a copy was sent to him for review in his capacity as a censor! The French Revolution put an end to his service in government, and he died in poverty in 1799.

William Wallace was born at Dysart, Fife, Scotland, in 1768. Apprenticed to a bookbinder in Edinburgh, he educated himself by reading science and mathematics books. After becoming a journeyman, he met Professors Robison and Playfair, who invited him to attend their lectures. He gave private lessons in mathematics and attended lectures for a year before obtaining a position at the academy at Perth in 1794. There he wrote and published mathematical articles; in 1803 he was appointed instructor of mathematics at the Royal Military College. Wallace wrote books on geometry and number theory and championed the teaching of Euclid's *Elements*. He also invented the eidograph, an improvement on the pantograph. In 1819, he was appointed professor of mathematics at Edinburgh, where he remained until his retirement in 1838. He died in 1843.

and flip the resulting triangles up to form a rectangle of length $z/2$ and height $2/z$, which is composed of three pieces. Then cut $\lfloor (z-1)/2 \rfloor$ rectangles of length 1 from the long rectangle, leaving a rectangle of length at least $\frac{1}{2}$ but less than $\frac{3}{2}$. This creates two more pieces for each unit-length rectangle. Cut the last rectangle with a P-slide when z is not an integer, or with the step technique otherwise. This adds three pieces when z is not an integer, and fewer otherwise, for a total of at most $6 + 2\lfloor (z-1)/2 \rfloor$ pieces, which is less than $z + 4$ when z is an integer or $\lfloor z \rfloor$ is even. The example in Figure 19.4 has $z = 4.47$ and uses eight pieces. The P-slide cut is shown with dashed edges.

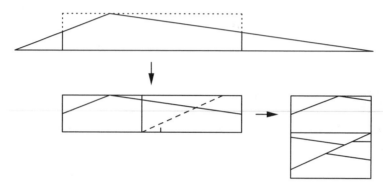

19.4: Irregular triangle to square: $\lfloor z \rfloor$ even

The second technique is just the TT-strip technique (see Chapter 12) restated: From the midpoint of the longest side, construct a line segment of length 1 to a point on the second longest side. Cut the triangle along this line segment, producing a smaller triangle and a quadrilateral, and flip the smaller triangle around. Draw lines perpendicular to the line segment at intervals of length 1, with one of the perpendiculars going through the midpoint of the short side of the original triangle. The smaller triangle has length $z/2$, and the quadrilateral has length at most $z/2 + 1$. At most $\lceil z/2 \rceil$ cuts separate the smaller triangle, giving $\lceil z/2 \rceil + 1$ pieces, and at most $\lceil z/2 + 1 \rceil$ cuts separate the quadrilateral, giving $\lceil z/2 \rceil + 2$ pieces. This gives a total of $2\lceil z/2 \rceil + 3$ pieces, which is less than $z + 4$ when $\lfloor z \rfloor$ is odd. The example in Figure 19.5 has $z = 5.75$ and uses nine pieces.

Of course, triangles are the simplest polygons. To dissect irregular polygons with more sides without using an overwhelming number of pieces, we choose polygons that are similar. The obvious problem is to dissect two similar polygons to a larger copy of the same. Harry Hart (1877) discovered two related dissections that apply to certain types of irregular polygons. The first type are those that can be circumscribed. By this we mean that there is a circle, the *circumcircle*, on which

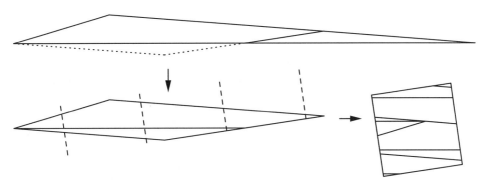

19.5: Irregular triangle to square: ⌊ z ⌋ odd

each vertex lies. We designate by {ⓟ} a *p*-sided polygon that can be circumscribed. The second type of polygons are those that can have an incircle inscribed within them. An *incircle* is a circle such that each side of the polygon is tangent to it. We designate by {p̊} a *p*-sided polygon that can have in incircle inscribed in it.

Hart's dissection for circumscribed polygons is in Figure 19.6. The larger of two pentagons is on the left, inside a circumcircle. For each side, Hart drew a line segment from its midpoint to the center of the circle. Since each side is a chord of the circle, each of these line segments is a perpendicular bisector of its side. From each vertex to the neighboring perpendicular bisector in a counterclockwise direction, he drew another line segment forming a right triangle whose legs are in the ratio of the pentagon sides. He then rotated the triangles out, around the midpoints of the sides, to give the arrangement of pieces in the middle of Figure 19.6. Finally, he shifted each pair of pieces out from the center and inserted the smaller of the two pentagons in the center, as shown on the right.

Hart's dissection does not apply to all ratios of side lengths for all polygons. When the portion of the circumcircle covered by a side of the polygon is greater

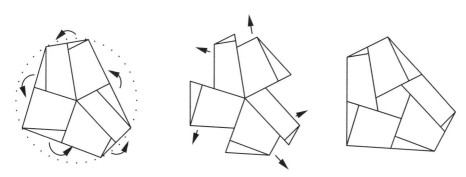

19.6: Hart's two circumscribed pentagons to one

than 90° for some side, then certain ratios are not possible. Hart's technique fails for equal-sized polygons, since an isosceles right triangle will extend beyond the end of the perpendicular bisector of the offending side. The largest ratio of side lengths of smaller to larger polygons is $\min_i\{\cot(\alpha_i/2)\}$, where α_i is the angle covered by the ith side.

Harry Hart was born in 1848 in Greenwich, England. He studied mathematics at Trinity College in Cambridge and earned a B.A. in 1871 and an M.A. in 1874. Appointed as an instructor in mathematics at the Royal Military Academy at Woolwich in 1873, he was promoted to professor of mathematics in 1884. From 1874 through 1885 he published some 25 papers on geometry and the geometry of motion. He died in 1920.

Hart's dissection for inscribed polygons is similar (Figure 19.7). The larger of two pentagons is on the left, with an incircle. From the center of the circle Hart drew line segments to the points of tangency of each of the five sides. From each vertex to the neighboring radius in a counterclockwise direction, he drew another line segment forming a right triangle whose legs have lengths in the ratio of side lengths of the pentagons. He then rotated the triangles around the vertices, to give the arrangement of pieces in the middle of Figure 19.7. Finally, he shifted each pair of pieces away from the center, to make room for the smaller of the two pentagons. Both of Hart's dissections use $2p + 1$ pieces, for figures with p sides.

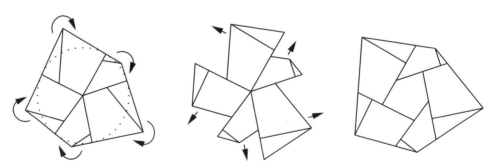

19.7: Hart's two inscribed pentagons to one

The second of Hart's dissections also does not apply to all ratios of side lengths for all polygons. If the polygon has an angle of less than 90°, then the largest ratio possible is $\min_i\{\tan(\beta_i/2)\}$, where β_i is the ith angle of the polygon.

An interesting feature of Hart's dissections is that there is a second way to cut the right triangles. For instance, in Figure 19.7 we could cut them on the clockwise side of each vertex, rather than the counterclockwise side. This suggests an approach for a two-to-two dissection of similar circumscribed or inscribed figures. Let's examine the inscribed pentagon in Figure 19.8, which we have copied from Figure 19.7. The dashed lines with longer dashes show the triangles cut in Figure 19.7, and the dashed lines with shorter dashes show another way to cut triangles, this time on the clockwise side of each vertex. Figure 19.8 is the basis for the two-to-two dissection shown in Figure 19.9, where each triangle is now the union of two triangles from Figure 19.8. This dissection, plus a similar one for circumscribed figures, uses $2p + 2$ pieces, for figures with p sides.

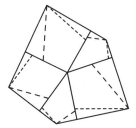

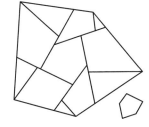

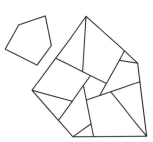

19.8: Both ways   **19.9:** Two pentagons to two

A century after Hart, David Collison (1980) found an economical dissection for a more general class of polygons. This class includes polygons in which at most one of the angles is greater than 180°. Collison based his dissection on dividing a p-sided polygon into $p - 2$ triangles and applying Harry Bradley's 5-piece two-triangles-to-one dissection to each triangle. Collison formed the triangles by cutting diagonals from the vertex of the largest angle to the nonneighboring vertices. He used two insights to reduce the number of pieces below $5(p - 2)$: First, he found that some pieces from one triangle dissection could be combined with the corresponding pieces in an adjacent triangle, if he reflected Bradley's dissection in every second triangle. Figure 19.10 shows the first step in Collison's dissection, with the boundary between Bradley's triangles shown with dotted lines in the large and small hexagons, and dotted or solid lines in the medium-sized hexagon. Collison did not cut any dotted lines, saving $2(p - 3)$ pieces.

Collison's second insight was that he could merge pairwise the triangles at the apex of the medium-sized hexagon if he left appropriate cavities for them to fit into. Figure 19.11 shows the final step in Collison's dissection, with dotted lines now indicating cuts that have been eliminated or shifted. Since for each pair of triangles Collison eliminated one cut and shifted another, he had a net saving of $\lfloor (p - 2)/2 \rfloor$ pieces. Thus the total number of pieces in Collison's dissection is $3p - 3 - \lfloor p/2 \rfloor$. The careful reader might observe that one triangle of a pair of triangles may be

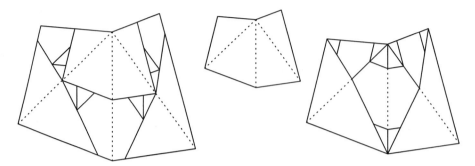

19.10: Collison's two irregular hexagons to one: first step

too thin to accommodate the shift. But the correction is simple: For each pair of triangles, use the broader of the two (measured by distance from the common side to each triangle's third vertex) as the one that receives the cut made into it. This has been done in Figure 19.11.

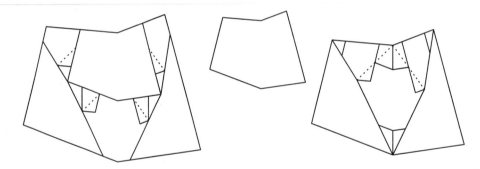

19.11: Collison's two irregular hexagons to one: final step

It is interesting to compare Collison's technique for irregular polygons with his technique for regular rational polygons, as in Figures 9.16 and 9.17. All that is missing here are the conversion of P-slides to steps and the further merging of triangles that is then possible. Collison did not make this connection explicit, but surely he must have understood it.

Let's also compare Collison's technique with Hart's. For small values of p, namely, $p < 7$, Collison's dissection uses fewer pieces than Hart's dissections. However, for $p > 8$, Hart's dissection wins out.

Lindgren (1964b) gave several examples of assemblies of four equal irregular polygons to one. One of these was of a rectangle that has a smaller rectangle cut from one of its corners. Using tessellation methods, Lindgren gave a 6-piece dissection, which was valid whenever both dimensions of the removed rectangle were at most half the size of the respective dimensions in the original rectangle. Robert Reid found that he needed no more than five pieces if he turned some over. The left half of Figure 19.12 shows his solution when the "notched rectangles" satisfy the constraint for Lindgren's 6-piece solution. If the length of the removed rectangle is greater than half the length of the original rectangle, or similarly for the widths, then the dissection of the left doesn't work, but Reid's dissection on the right does. When the rectangle removed has both dimensions more than half of the original rectangle's corresponding dimensions, then six pieces seem necessary.

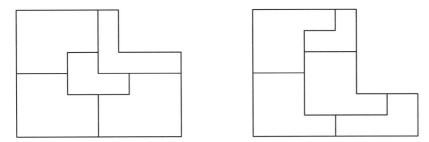

19.12: Reid's four notched rectangles to one: two cases

CHAPTER 20

On to Solids

"Who of us would not be glad to lift the veil behind which the future lies hidden; to cast a glance at the next advances of our science and at the secrets of its development during future centuries?" So, stirringly, began David Hilbert at the beginning of the twentieth century. His wide-ranging list of twenty-three problems, heralded before the International Congress of Mathematicians in Paris, has inspired a stampede of research during this, the first of his future centuries. And the exciting news was that the third problem on his list concerned dissections!

It sounded wonderful, but there was a flip side: First, Hilbert proposed a negative result, challenging mathematicians to prove the impossibility of certain three-dimensional dissections. Second, the problem was solved before the congress even got under way. Third, the problem has since been viewed as one motivated primarily by pedagogical concerns. So much for geometric dissection's brief membership in the vanguard of twentieth-century mathematics!

What a letdown! And although three-dimensional dissections were only briefly in that brightest of spotlights, we shall find that there have been some nice examples produced even in the shadows. We start our story a half century before dissection's fifteen minutes of fame. In an exchange of letters between Carl Friedrich Gauss and Christian Ludwig Gerling in 1844, Gauss had lamented that the only known procedure for proving the equality of the volume of a polyhedron with the volume of its mirror image was a method of exhaustion. However, in his return letter, Gerling

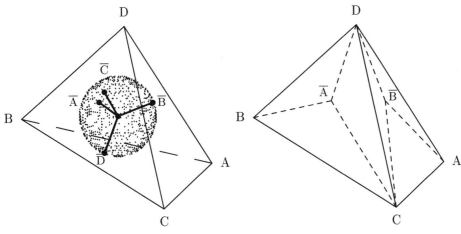

20.1: Inscribed sphere   **20.2:** Cutting tetrahedron

described a simple proof that established that an irregular tetrahedron has the same volume as its mirror image. Since a polyhedron can be split into tetrahedra, the result followed. Gerling gave a simple dissection of the tetrahedron into twelve pieces that can be reassembled into the mirror image. His dissection uses a circumscribed sphere, but "negative pieces" arise when the center of the sphere falls outside the tetrahedron.

Juel (1903) gave a related 12-piece dissection that uses the inscribed sphere. Børge Jessen (1968) combined pairs of pieces in Juel's dissection to give a 6-piece dissection (Figure 20.3). It is easy to produce this three-dimensional analogue of the dissection in Figure 3.9: First, inscribe a sphere in the tetrahedron (Figure 20.1). Large dots represent the center of the sphere and the points of tangency. The point of tangency with the face opposite vertex A is $\overline{A}$, with $\overline{B}$, $\overline{C}$, and $\overline{D}$ similarly named. Thick lines indicate the radii from the center to these points.

Next, cut the tetrahedron into six pieces. Base each piece on one edge of the tetrahedron, the points of tangency on the two faces that contain that edge in their boundaries, and the center of the inscribed sphere. The dashed lines in Figure 20.2 indicate the position of the cuts when the six pieces of the tetrahedron are assembled. In Figure 20.3 a large dot indicates the vertex in each piece that corresponds to the center, and each remaining vertex is labeled with the point in Figure 20.1 with which it corresponds. We can move the pieces in Figure 20.3 by translation

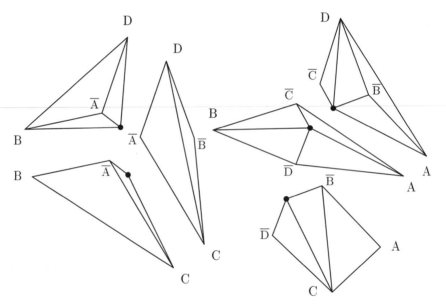

20.3: Irregular tetrahedron to its mirror image

back into the tetrahedron in Figure 20.2. Each piece has five vertices, some obscured from our view. Further, each piece has a plane of symmetry that contains its edge from the tetrahedron plus the center of the inscribed sphere. Since the mirror image tetrahedron will have a dissection to the same set of six pieces, we get the desired dissection. There is a simple generalization to any polyhedron in which a sphere can be inscribed, and the number of pieces equals the number of edges in the polyhedron.

Five years before Hilbert announced his list of problems, M. J. M. (Micaiah John Muller) Hill (1896) had taken a constructive approach. Hill, a professor of mathematics at the University College in London, had identified three infinite classes of tetrahedra whose volumes could be determined without resort to a method of limits. Any tetrahedron in his classes can be converted to a cube by dissection, and hence to any other tetrahedron in his classes.

We can derive Hill's first type of tetrahedron from a *rhombohedron*, which is a polyhedron with six faces, each of which is a rhombus. (We can produce the vertex-edge skeleton of a rhombohedron from the vertex-edge skeleton of a cube by moving two diagonally opposite vertices A and D either toward each other or away from each other. With the edges rigid and the vertices like universal joints, the skeleton will expand or contract like an accordion.) The rhombohedron decomposes into six tetrahedra, three identical and three mirror images. Each tetrahedron will contain the two vertices A and D, and two other vertices B and C such that AB, BC, and CD are edges of the rhombohedron. For a parameter r, with $0 < r < \sqrt{2}$, and a scaling coefficient a, the Hill tetrahedron of the first type will have edges AB, BC, and CD of length $a\sqrt{1 + r^2}$; edges AC and BD of length $2a$; and edge AD of length $a\sqrt{9 - 3r^2}$. Actually, this first type of tetrahedron, and the decomposition of a rhombohedron into six such tetrahedra, were described already by René Haüy (1801). Thus these could rightly be called Haüy-tetrahedra, which Hill then rediscovered.

Hill's second and third types of tetrahedron can easily be derived from the first. Let E be the midpoint of edge AD in a Hill tetrahedron of the first type. Then the tetrahedron with vertices A, B, C, and E is a Hill tetrahedron of the second type. Let F be the midpoint of edge BC in a Hill tetrahedron of the first type. Then the tetrahedron with vertices A, B, D, and F is a Hill tetrahedron of the third type.

At the time when Hill defined his tetrahedra, they seemed to be the exception, rather than the rule, with respect to dissection to a cube. For example, no one knew at that time how to dissect a regular tetrahedron to a cube in a finite number of pieces. (In fact, it can't be done.) Bricard (1896) had already attempted a proof of the conjecture that was embodied in Hilbert's third problem. Although incorrect, the attempted proof contained crucial elements of the eventual proof by Dehn (1900). The proof involves defining for each polyhedron what is now called its *Dehn invariant*, a sum of terms each of which is the length of an edge times a certain function of the angle of the two faces meeting at that edge. Dehn completed the proof by identifying a pair of polyhedra such that their Dehn invariants are different. Jean-Pierre Sydler (1965) showed that the condition of Dehn invariants was not only necessary but also sufficient. In other words, one polyhedron is dissectable to another if and only if for all appropriate functions, their Dehn invariants are equal. Disappointingly, no regular polyhedron is dissectable to any other regular polyhedron.

M. J. M. Hill did not supply dissections to illustrate the decomposability of any of his tetrahedra to a cube. The Swiss did this more than a half-century later. First,

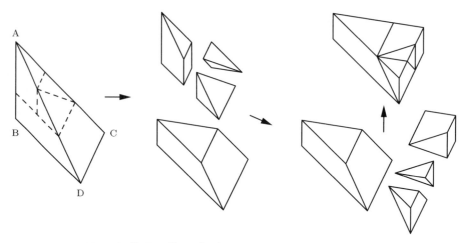

20.4: Sydler's Hill tetrahedron to isosceles triangular prism

Sydler (1956) demonstrated how to dissect a particular Hill tetrahedron of the first type to a right prism whose base is an isosceles right triangle. Once we have a prism, it is easy to convert the figure to a rectangular block, and hence to a cube. For this example, $r = 1$, so that the dihedral angles at edges AC, BC, and BD are 90°; that implies that edges AB and CD are perpendicular to edge BC. The corresponding rhombohedron is precisely the cube. Sydler cut this tetrahedron into four pieces that rearrange to form a prism, where the top face of the prism is an isosceles right triangle (Figure 20.4). An analogous dissection applies for any tetrahedron of Hill's first type, giving a prism that is not a right prism. It is easy to convert such a prism to a right prism.

Remarkably, there is an even nicer dissection of that tetrahedron to a different triangular prism. Philipp Schöbi (1985), and independently Anton Hanegraaf, found

Jean-Pierre Sydler was born in Neuenburg, Switzerland, in 1921. At the Eidgenössische Technische Hochschule (ETH), Zurich, he earned a graduate diploma in mathematics in 1943, and a Dr. sc. math. in 1946. In 1950 he left his position in the mathematics department to join the staff of the library at ETH. There he continued to publish mathemat- ical papers in his spare time. In 1966 he was awarded the gold medal of the Royal Danish Academy of Sciences and Letters for resolving Hilbert's third problem. Becoming director of the library in 1963, Sydler pioneered the use of automation in the library system. He served as direc- tor until his retirement in 1986. He died in 1988.

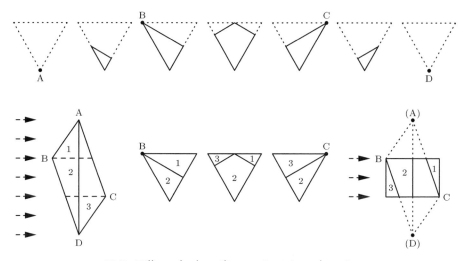

20.5: Hill tetrahedron (first type) to triangular prism

an elegant 3-piece dissection of any Hill tetrahedron of the first type to a triangular prism, where the triangle is equilateral.

Figure 20.5 orients the tetrahedron so that vertex A is directly above vertex D. In this orientation, the projected outline of the tetrahedron, when viewed from above, is an equilateral triangle. Cross sections of Hill's tetrahedron appear in the top row of this figure, with the projected outline of the tetrahedron shown with dotted lines. The bottom row illustrates the dissection itself, with front views of the tetrahedron and prism given on the extreme left and right, respectively. The dissection is exceptionally simple, with just two horizontal slices made, one through vertex B and the other through vertex C. The pieces may be hinged, with pieces 1 and 2 hinged at the level of B, and pieces 2 and 3 hinged at the level of C. The arrows indicate the levels at which cross sections are viewed. The middle of the bottom row gives three cross sections of the prism, at the levels indicated. Hanegraaf also noted that for one of the tetrahedra of Hill's first type, the resulting thickness of the triangular prism exactly equals the side length of a cube of the same volume. In this case, we can project the triangle-to-square dissection through the height of the prism. This gives a 9-piece fully hinged dissection of a particular tetrahedron of Hill's first type to a cube.

Schöbi, a mathematics student at ETH-Zurich, who received his doctorate in 1990, also extended the idea behind the dissection of Figure 20.5 to the Hill tetrahedra of the second and third type: He (1985) gave 4-piece and 6-piece dissections, respectively, for these figures to triangular prisms. Schöbi's dissection for the third

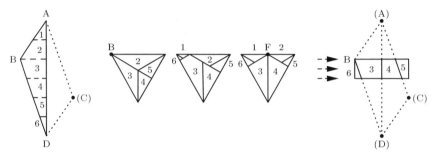

20.6: Hill tetrahedron (third type) to triangular prism

type is in Figure 20.6. An example of Hill's tetrahedra of the third type is on the left, with vertex A directly above vertex D, and with the outline of the tetrahedron of the first type from which it is derived shown with dotted lines. Schöbi sliced the tetrahedron by five horizontal slices into six pieces. Vertex F is at the rear of the right-hand boundary between pieces 3 and 4. The resulting prism is on the right, with three cross sections shown in the middle. The resulting prism is one-half the height of the prism in Figure 20.5.

Puzzle 20.1: Find a 4-piece dissection of Hill's tetrahedron of the second type to a triangular prism.

The preceding examples of tetrahedral dissections arose in response to concerns about demonstrating equivalence of volume with dissection methods. An assumption that underlies the viewpoint of Hill, Hilbert, and Dehn is that dissecting a rectangular solid to a cube is easy. Surprisingly, there seems to have been no explicit mention of such a dissection until William Fitch Cheney (1933) gave an 8-piece dissection of a $(2 \times 1 \times 1)$-rectangular block to a cube. The resulting cube has each dimension equal to $r = \sqrt[3]{2}$. Cheney first converted the $(2 \times 1 \times 1)$-block to an $(r \times r^2 \times 1)$-block, using a P-strip on a face that is a (2×1)-rectangle, to convert it to an $(r \times r^2)$-rectangle. He projected the cuts on this face into the third dimension along planes perpendicular to that face, creating three pieces. Cheney then converted the $(r \times r^2 \times 1)$-block to the cube by a similar operation on a face that is a $(r^2 \times 1)$-rectangle. The cuts from the second operation intersect those of the first, giving a total of eight pieces.

Before Cheney's solution could appear in print, Albert Wheeler (1935) discovered a 7-piece solution. The improved dissection had the same strategy of converting the original block to an $(r \times r^2 \times 1)$-block and then to a cube. However, Wheeler used a P-slide in place of each P-strip. By orienting the second P-slide appropriately,

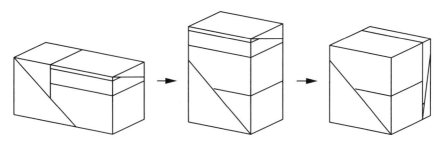

20.7: Hanegraaf's (2 x 1 x 1)-rectangular block to cube

he saved a piece. Care in avoiding intersecting cuts is a key insight, which, it turns out, Wheeler did not exploit fully, for Wheeler's dissection is not minimal, either!

A 6-piece dissection by Anton Hanegraaf (1989) is in Figure 20.7. Hanegraaf began as Wheeler had, by using a P-slide to form the $(r \times r^2 \times 1)$-block. In the second step, Hanegraaf avoided crossing previous cuts by slicing the $(r \times r^2 \times 1)$-block into an $(r \times (r^2-r) \times 1)$-block and an $(r \times r \times 1)$-block. He then used a P-slide to convert the $(r \times (r^2-r) \times 1)$-block to a $(r \times (r-1) \times r)$-block. He rotated the resulting block and placed it next to the $(r \times r \times 1)$-block to give the cube. Although Hanegraaf saved a piece, his dissection is not translational, as was Wheeler's. Hanegraaf's approach works on rectangular blocks spanning a range of different dimensions.

If we do not mind using more pieces, then it is possible to create a dissection that is cyclicly hinged. This was first realized by Michael Goldberg (1966), who used two applications of Harry Lindgren's Q-slide technique. Goldberg claimed that his dissection had ten pieces, but, as Hanegraaf pointed out, it actually had twelve pieces, two of which were very thin triangular prisms. But why quibble, because Hanegraaf (1989) gave a cyclicly hinged dissection that has just seven pieces!

Hanegraaf used the same strategy as Cheney and Wheeler, converting the original block to an $(r \times r^2 \times 1)$-block and then to a cube, but employed his T-slide technique (Figure 11.31). The T-slide is particularly nice when the trapezoid is

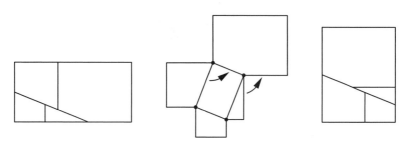

20.8: Hanegraaf's hinged slide

237

a parallelogram, since there is no cut that crosses the parallelogram along the long dimension. A T-slide takes a (2×1)-rectangle to an $(r \times r^2)$-rectangle in Figure 20.8. A second illustration in Figure 20.9 converts a $(1 \times r^2)$-rectangle to a square.

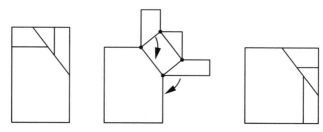

20.9: A second hinged slide by Hanegraaf

Hanegraaf applied the dissections in Figures 20.8 and 20.9 to the appropriate faces of blocks in the solid dissection, with the cuts projecting through the third dimension, as in Cheney's and Wheeler's dissections. The resulting 7-piece dissection, a variant of Hanegraaf's, is illustrated in Figure 20.10. The figure shows the $(2 \times 1 \times 1)$-block on the top left and the cube on the top right, with intermediate steps in the pair of hinged transformations in between.

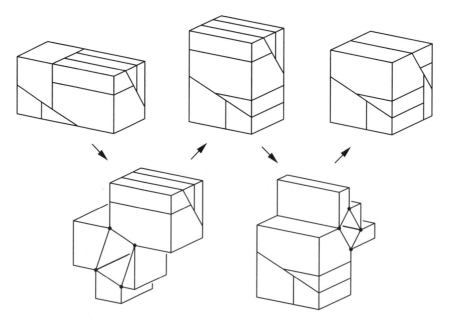

20.10: Hanegraaf's hinged $(2 \times 1 \times 1)$-rectangular block to cube

Suppose that we replace the $(2 \times 1 \times 1)$-block by two equal cubes. How many more pieces does it take to dissect the pairs of cubes, rather than the block, to a larger cube? If we chop the $(2 \times 1 \times 1)$-block of Figure 20.7 in half, this gives us an 11-piece dissection. Hanegraaf (1989) found a 10-piece solution, then hid it within Figure 20.15. Can the reader find it?

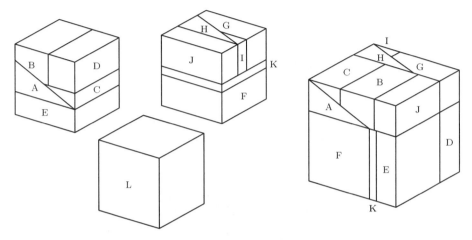

20.11: Hanegraaf's three cubes to one

Hanegraaf next turned to three cubes to one. His clever 12-piece dissection is shown in Figure 20.11. Let $s = \sqrt[3]{3}$ be the size of each dimension of the resulting cube. The s-cube is assembled from an $(s \times (s-1) \times (s-1))$-block (pieces G, H, and I) and an $(s \times 1 \times (s-1))$-block (pieces A, B, C, and J) on the top, and an $((s-1) \times (s-1) \times s)$-block (piece D), an $((s-1) \times 1 \times 1)$-block (pieces E and K), an $((s-1) \times 1 \times 1)$-block (piece F), and a $(1 \times 1 \times 1)$-block (piece L) on the bottom. Piece L is obscured in the s-cube by the other pieces in Figure 20.11. Hanegraaf used P-slides in two ways: His rearrangement of pieces G, H, and I is standard, but he cut pieces A, B, and C so that they fit around piece D in the 1-cube and also fill in around piece J in the s-cube.

Let's shift from equal to unequal cubes, asking for a physical realization of solutions to $x^3 + y^3 = z^3$, with $x < y$. Anton Hanegraaf's detailed study of these problems identified a number of cases in terms of the relative sizes of x and y. It used two general approaches. The first left the x-cube intact and converted the y-cube to three blocks: an $(x \times x \times (z-x))$-block, an $(x \times (z-x) \times z)$-block, and a $((z-x) \times z \times z)$-block. If $x < y(\sqrt{31/3} - 1)/4 \approx 0.5536y$, then eleven pieces result. We see this case in Figure 20.12, where the x-cube is piece A; the first block above

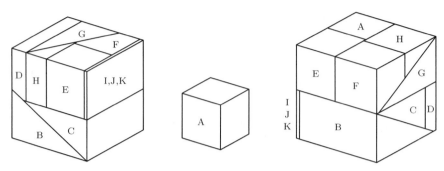

20.12: Cubes for $x^3 + y^3 = z^3$, with $x/y = 1/2$

is piece E; the second consists of pieces F, G, and H; and the third the remaining pieces B, C, D, I, J, and K. The last three pieces effect a P-slide from a $((z-x) \times (z(z-y)/y) \times y)$-block in the y-cube to a $((z-x) \times (z-y) \times z)$-block in the z-cube. Our example demonstrates the case when $x/y = 1/2$. Unfortunately, as this ratio gets smaller, pieces I, J, and K get extremely narrow. If $y(\sqrt{31/3} - 1)/4 < x < y(\sqrt{5} - 1)/2 \approx 0.618y$, then twelve pieces result.

The second approach converted the x-cube to a $(z \times z \times (x^3/z^2))$-block, converted the y-cube to a $(z \times z \times (y^3/z^2))$-block, and then set these blocks side by side. This gave a 12-piece dissection for $x > y \sqrt[3]{1/(2\sqrt{2} - 1)}$. For $y(\sqrt{5} - 1)/2 < x < y \sqrt[3]{1/(2\sqrt{2} - 1)}$, either approach used thirteen pieces. In all cases, Hanegraaf used the P-slide in a fashion similar to that in three equal cubes to one. He also surveyed those cases for which a step can replace one or more of the P-slides.

Besides cubes, truncated octahedra are not very hard to dissect, in part because they honeycomb three dimensions. A direct consequence of this honeycomb is a 6-piece dissection of two truncated octahedra to one cube (Figure 20.13), which both David Paterson (1988) and Anton Hanegraaf noted. They cut the first truncated octahedron into four equal pieces by a horizontal slice and a vertical slice. Then they

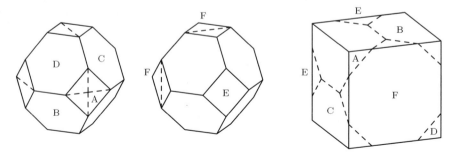

20.13: Two truncated octahedra to cube

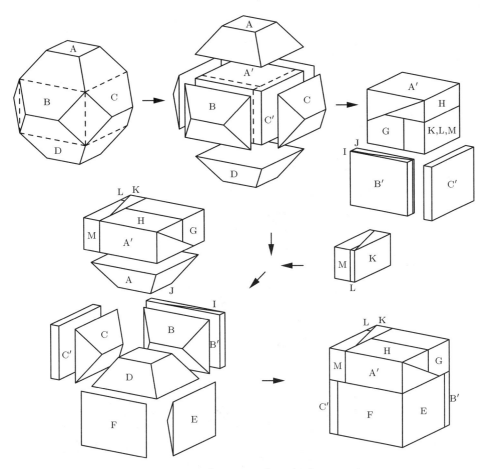

20.14: Hanegraaf's truncated octahedron to cube

cut the second truncated octahedron into two equal pieces by a slice perpendicular to the planes of the other two slices. The resulting dissection is translational, with reflection and 4-fold rotational symmetry.

More challenging is dissecting one truncated octahedron to one cube. Paterson (1988) gave a 17-piece dissection, in which he used a generalization of the P-strip technique in three dimensions. However, Hanegraaf found a 13-piece dissection, based once again on the P-slide technique.

Hanegraaf's dissection is illustrated in Figure 20.14. First, he cut away the top and bottom (pieces A and D), and also the four sides (pieces C, D, E, and F), as indicated by the dashed edges on the upper left. These cuts leave a rectangular block, and the six pieces form an identical rectangular block, so that the remaining work combines these two blocks into a cube. Then he divided the solid rectangular block into three blocks by the dashed lines in the second part of the figure. If an edge of the truncated octahedron is of length 1, then piece C′ is $(2\sqrt[6]{2} - 2)$ wide. A P-slide makes the second block the same width and changes its length to $2\sqrt[6]{2}$. The P-slide creates pieces B′, I, and J. Hanegraaf cut the remaining block into six pieces (A′, G, H, K, L, M) via two P-slides, which together form a $(2\sqrt[6]{2} - \sqrt{2})$-wide slice of the resulting cube. We see an exploded view of the cube and the assembled cube at the bottom of the figure. At first it appears that Hanegraaf used sixteen pieces, but the reader can verify that A and A′ need not be separated, and similarly for B and B′, and for C and C′. So Hanegraaf's dissection uses only thirteen pieces.

The rhombic dodecahedron is another three-dimensional figure that is not hard to dissect. In Chapter 2, we discussed how to dissect a rhombic dodecahedron into seven pieces that fit together to form two cubes. Of course, we can then dissect it to one cube, and Anton Hanegraaf found a nifty 13-piece dissection for this case. Hanegraaf's dissection is similar to his dissection of a truncated octahedron to a cube and is illustrated in Figure 20.15. He cut four of the six caps (pieces B, C, D, and F) off the rhombic dodecahedron and arranged them to form two-thirds of a small cube. The remaining two caps, extended with rectangular blocks, form pieces A and E, which fill out the small cube and extend it to fill out a portion of the

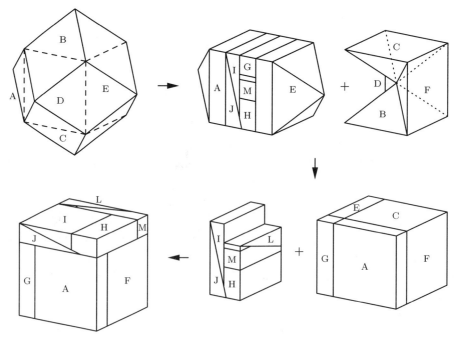

20.15: Hanegraaf's rhombic dodecahedron to a cube

desired cube. He cut the remainder of the rhombic dodecahedron into seven pieces (G through M), using two P-slides, to fill out the remaining volume of the desired large cube. The small unlabeled wedge in several of the objects is piece K.

Two tetrahedra plus an octahedron form an element that can honeycomb 3-space. Thus there is a dissection of two tetrahedra and an octahedron to a cube, even though neither alone has a dissection to a cube. Duilio Carpitella (1996) gave a symmetrical 16-piece dissection. The honeycomb element is the parallelopiped shown on the top right in Figure 20.16. In a fashion analogous to forming a strip, we can line up copies of this element, one after another, to form a *post*, whose cross section is a parallelogram.

We can then place posts flush against each other to form a *wall* of uniform thickness. We can shift the posts relative to each other, in the same way that we shift strips when we tessellate the plane. We can also create posts and walls when using a cube as a honeycomb element, but crossposing the two posts is difficult to visualize in three dimensions. Carpitella simplified the approach by slicing the cube so that the two resulting pieces form a parallelopiped whose height equals the thickness of the tetrahedron–octahedron wall. The sliced cube and the resulting parallelopiped appear in the middle right and lower right of Figure 20.16. The dotted lines give the outline of a rectangular box in which the parallelopiped fits. We can now use this parallelopiped as a honeycomb element to form a post, and then a wall.

To produce the dissection, crosspose the tetrahedron–octahedron post with the parallelopiped post, while constraining the walls induced by the two posts to

---

Anton Hanegraaf was born in 1931 in Diessen, the Netherlands. He completed studies in civil engineering at the technological college in 's-Hertogenbosch in 1952. After serving two years of military service in the Engineering Corps, he worked for three years as a structural designer for an architectural firm, and for several more as a free-lance structural engineering consultant. From 1962 until his retirement in 1992, he was an employee of ABT, a firm of structural engineering consultants. His work involved both practical and theoretical research, the development of structural analysis methods, the study of the behavior of structural materials, the study of structural shapes, and the study of morphological aspects of space enclosure.

Anton's recreational pursuits include mechanical and mathematical puzzles, mathematical art, and listening to music and reading a wide variety of books. His mathematical interests include mathematical models, mechanisms and linkages, tilings, polyhedra, and sliding piece and rotational puzzles. He was editor of *Cubism for Fun* from 1987 to 1993. He plans to publish five more booklets in his series *Polyhedral Dissections*.

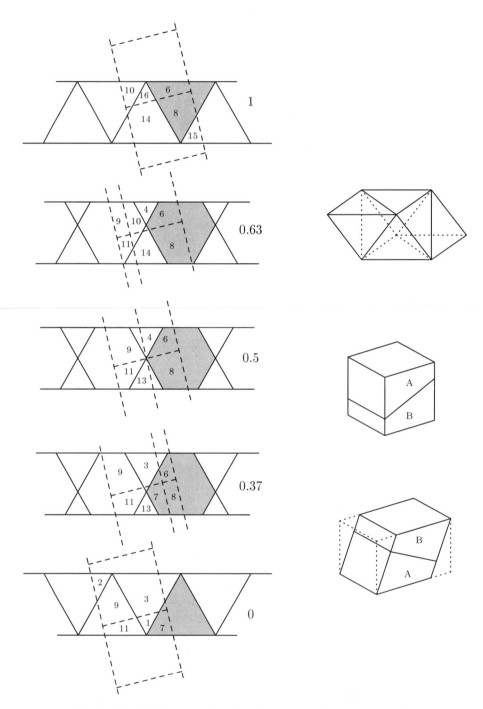

20.16: Carpitella's one octahedron plus two tetrahedra to one cube

coincide. Since this is still challenging to visualize, we rely on the cross sections shown on the left in Figure 20.16. Each of the five cross sections gives the crossposition at a different height from the bottom of the wall: 1, 0.63, 0.5, 0.37, 0, where 1 is at the top, 0.5 is at the middle, and 0 is at the bottom. One of the octahedra in each of the cross sections is shaded.

We can number the sixteen pieces in the dissection, with the first four pieces forming one tetrahedron, the last four forming the second tetrahedron, and each piece and its mirror image having numbers that sum to 17. Some of the pieces are either too small or too thin to be labeled in each cross section. At height 1, the unlabeled triangular piece adjacent to pieces 10 and 14 is piece 12. At height 0.63, the unlabeled pieces are 16, 12, and 13, in clockwise order around piece 14. At height 0.5, the unlabeled pieces adjacent to pieces 4 and 13 are, respectively, 3 and 14. We can infer unshown labels at levels 0.37 and 0 in a fashion similar to those on levels 0.63 and 1.

Carpitella (1996) investigated a variety of three-dimensional dissections that rely on three-dimensional analogues of plain-strip dissections. When necessary, a three-dimensional element for one strip may be sliced through a plane to give an appropriate thickness in the resulting element. Carpitella employed this approach to give 14-piece symmetrical dissections of a rhombic dodecahedron to a cube, and of a truncated octahedron to a cube. He also found a 28-piece dissection of a rhombic dodecahedron to a truncated octahedron.

Our excursion into three-dimensional dissections has been subject to rather severe limitations on which dissections are possible. However, Gavin Theobald posed a natural (and in hindsight obvious) antidote: Dissect the surfaces instead! This leads to many surprising results, one of which we will see here. Theobald discovered a number of 2-piece dissections, based on the flexibility of cutting and opening out the surfaces of three-dimensional figures. (He also discovered some other surface dissections that used more than two pieces.)

Duilio Carpitella was born in Trapani, Sicily, Italy, in 1957. After attending the First Artistic High School in Rome, he enrolled at the Faculty of Architecture at the University of Rome ("La Sapienza"). He completed his studies in 1983 and worked in an architectural office for four years, before opening a studio in Rome with two colleagues to work with illustration. A major interest of Carpitella's has been painting; his work derives from an analysis of M. C. Escher's work on tessellations. He has participated in exhibitions of work inspired by Escher and on mathematical imagery. He also enjoys inventing games.

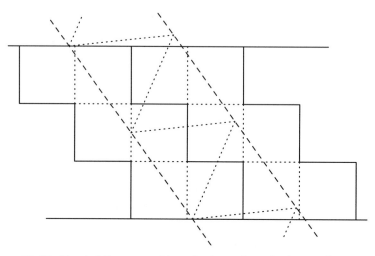

20.17: Theobald's crossposition of cube and tetrahedron surfaces

Theobald cut the surface of a tetrahedron along two opposite edges to give a band, then with one more cut gave a parallelogram. In addition, Theobald unfolded the surface of a cube in a zigzag fashion to give a P-strip element. The associated crossposition is in Figure 20.17, where the solid lines show the cube's P-strip element, the dashed edges show the boundary of the tetrahedron P-strip, and the dotted lines show the folds. A view of the 2-piece dissection is in Figure 20.18. In the cube, solid lines indicate the cuts, dashed lines the edges that are not cut, and dotted lines the folds from the tetrahedron. We can use the same convention in the tetrahedron, except that there the dotted lines represent folds from the cube.

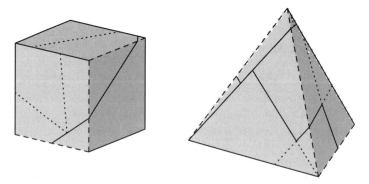

20.18: Theobald's surface of cube to surface of tetrahedron

CHAPTER 21

Cubes Rationalized

In the twilight of his career at Cambridge University, Herbert W. Richmond combined his lifelong interests in geometry and number theory to pose one of the more natural of puzzles in solid dissections: Dissect cubes of edge lengths 3, 4, and 5 to give a cube of edge length 6. The venerable octogenarian produced a dissection with some appealing symmetries, but it used twelve pieces – a rather generous number.

Six years later, John Leech, then a student at Cambridge, latched onto the problem. He challenged the readers of Eureka, *the Cambridge undergraduate math journal, to find a dissection that used at most ten pieces, an improvement of two over Richmond's solution. And within a year, the charge of that young generation was complete. Roger Wheeler, also an undergraduate at Cambridge, responded with a dissection that reduced the number of pieces even further, again by two, down to eight!*

What colossal impertinence in that young generation at Cambridge! Or was it just a colossal impatience, given the few solid dissection problems that had been posed and solved? Surely it was implacability in the face of problems that seemed so much more difficult than their two-dimensional analogues. It had been a natural step for Richmond (1943) to pose the dissection for cubes realizing $3^3 + 4^3 + 5^3 = 6^3$. And considering that there were no previous dissections for any solution to $x^3 + y^3 + z^3 = w^3$, twelve pieces might have seemed a veritable bargain. That the dissection by Richmond (1944) was subsequently beaten by four pieces suggests that rational dissections in three dimensions are considerably more difficult than those in two dimensions. It also suggests that $3^3 + 4^3 + 5^3 = 6^3$

may be an especially economical case, much as $1^2 + 2^2 + 2^2 = 3^2$ is for two dimensions.

Herbert William Richmond was born in 1863 at Tottenham, Middlesex, England. He received a B.A. in mathematics at King's College, Cambridge, in 1885 and an M.A. in 1889. He received a fellowship at King's in 1888, became a college lecturer in 1891, and was also a university lecturer before his retirement in 1928. Richmond's main focus was on algebraic geometry, and even his work on number theory, on the sums of cubes of three rational numbers, promoted a geometric point of view.

During the First World War, Richmond performed both routine computation and research on the ballistics of shell flight. Subsequently, he wrote a textbook on anti-aircraft gunnery. With regard to his mathematical work, Richmond wrote in 1946 that he "derived the pleasure that comes from a hobby rather than a business," and he continued his research until his death in 1948.

There are infinitely many integral solutions to $x^3 + y^3 + z^3 = w^3$, and formulas exist for generating these solutions. But the formulas don't seem as simple as the formulas for squares. Furthermore, general methods that apply to classes of solutions have not yet been discovered. Consequently, we shall see economical dissections for only a handful of special cases. In none of these cases do we have a minimal dissection that is translational. Moreover, the pieces that are rotated are rotated in different directions. Until an even more implacable generation comes to our rescue, we will have to rationalize our plight and treasure the few gems that we have.

Martin Gardner (1986) identified John Leech as the person who posed the problem of finding a 10-piece dissection in (Anonymous 1950). Roger Wheeler's 8-piece surprise (Figure 21.1) appeared in (Anonymous 1951). Wheeler left the 3-cube (piece A) uncut, cut the 4-cube into pieces B and C, and cut the 5-cube into pieces D through H. Pieces F and H are shown in Figure 21.2. The dissection appears to have been the result of inspired trial and error.

There are a number of other dissections for $3^3 + 4^3 + 5^3 = 6^3$ that have interesting features. Gardner (1986) described an 8-piece dissection in which the 4-cube remains intact, which was found by J. H. Thewlis of Argyll, Scotland, and simplified by Thomas H. O'Beirne. He also reported an 8-piece dissection in which the 5-cube was left intact, found by Emmet J. Duffy of Oak Park, Illinois. Finally Gardner reproduced a 9-piece dissection by Thomas H. O'Beirne in which each piece is a rectangular solid.

One formula is adequate for generating all integer solutions to both $x^3 + y^3 = z^3 + w^3$ and $x^3 + y^3 + z^3 = w^3$, if we allow the unknowns to be positive or negative. The following method for solutions to $x^3 + y^3 + z^3 = w^3$ is due to the

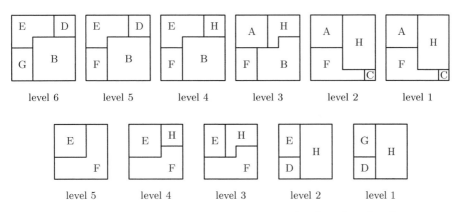

level 6   level 5   level 4   level 3   level 2   level 1

level 5   level 4   level 3   level 2   level 1

21.1: 6-cube and 5-cube in Wheeler's cubes for $3^3 + 4^3 + 5^3 = 6^3$

nineteenth-century mathematician J. P. M. Binet. For any integers m and n, we take

$$x = 1 - (m - 3n)(m^2 + 3n^2), \qquad y = (m + 3n)(m^2 + 3n^2) - 1,$$

$$z = (m^2 + 3n^2)^2 - (m + 3n), \quad \text{and} \quad w = (m^2 + 3n^2)^2 - (m - 3n)$$

The solutions are often not in reduced form, however. Thus $m = 0$ and $n = 1$ give $10^3 + 8^3 + 6^3 = 12^3$, and $m = 1$ and $n = 1$ give $9^3 + 15^3 + 12^3 = 18^3$. Dividing each base in the first by 2, or in the second by 3, gives our familiar $3^3 + 4^3 + 5^3 = 6^3$. If we take $m = -2$ and $n = 1$, we get $36^3 + 6^3 + 48^3 = 54^3$. Dividing by 6 gives $6^3 + 1^3 + 8^3 = 9^3$, which we shall discuss shortly.

Any rational dissection of three cubes to one requires at least eight pieces, since each corner of the w-cube must be in a separate piece. Thus Wheeler's dissection uses a minimum number of pieces. A rational dissection of two cubes to two other

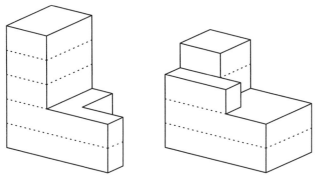

21.2: Pieces F and H in Wheeler's cubes for $3^3 + 4^3 + 5^3 = 6^3$

249

cubes requires nine pieces: When x is the largest of x, y, z, w, then each corner of the x-cube must be in a different piece and the y-cube must contribute at least one more piece.

With impertinence, impatience, or whatever, the English have cornered the market on these problems. James H. Cadwell, himself a Cambridge graduate, discovered several ingenious dissections in the 1960s. For cubes realizing $1^3 + 6^3 + 8^3 = 9^3$, Cadwell (1964) gave a 9-piece dissection. With colossal improvisation, Robert Reid, an expatriate Englishman in Peru, later found the 8-piece dissection given in

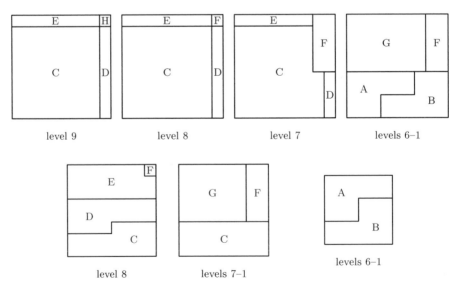

21.3: 9-, 8-, and 6-cube in Reid's cubes for $1^3 + 6^3 + 8^3 = 9^3$

Figure 21.3. He left the 1-cube (piece H) uncut and cut the 6-cube with the step technique into two pieces (A and B) that rearrange into a $(6 \times 4 \times 9)$-block. Reid handled the 8-cube cleverly: It is as though Reid first sliced off a $(1 \times 1 \times 8)$-block and cut the remainder into a $(5 \times 6 \times 7)$-block (piece G), a $(2 \times 5 \times 7)$-block (the basis for piece F), and a $(3 \times 8 \times 7)$-block (the basis for piece C). Then he cut the $(1 \times 1 \times 8)$-block into four parts: pieces E and D and two others that Reid reattached to pieces F and C. Pieces A, B, G, and F fill the first six levels of the 9-cube, and pieces C, D, E, H, and F (again) fill the remaining three levels.

Not to chance being twice improved upon, Cadwell (1964) gave a 9-piece (and thus minimal) dissection for cubes realizing $12^3 + 1^3 = 10^3 + 9^3$. We see a minor variation of this dissection in Figure 21.5. Cadwell left the 9-cube (piece A). He cut the 1-cube (piece I) and a $(7 \times 7 \times 9)$-block out of the 10-cube, leaving piece B. Then he cut the $(7 \times 7 \times 9)$-block into a $(2 \times 7 \times 9)$-block (piece C), two $(2 \times 3 \times 9)$-blocks (pieces D and E), and three additional pieces (F, G, and H). Cadwell's nice insight was to cut the $(7 \times 7 \times 9)$-block out of the 10-cube, so that the 9-cube would fit into the resulting cavity when he formed the 12-cube. Also, Cadwell used the step technique on pieces G and H to go from length 9 in the 10-cube to length 12 in the 12-cube. A perspective view of these pieces is shown in Figure 21.4.

It is one of those imponderable facts that 1729, which is the sum of $10^3 + 9^3$, has three factors: 7, 13, and 19. After discovering his dissection for $12^3 + 1^3 = 10^3 + 9^3$, James H. Cadwell was challenged by a colleague to dissect a $(7 \times 13 \times 19)$-block into

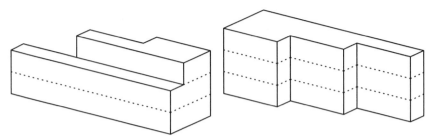

21.4: Pieces G and H in Cadwell's cubes for $12^3 + 1^3 = 10^3 + 9^3$

pieces that would form either pair of cubes. Cadwell (1970) found an ingenious three-way dissection to perform this conversion. The dissection in Figure 21.6 uses only twelve pieces.

Cadwell formed the $(7 \times 13 \times 19)$-block from a $(7 \times 13 \times 9)$-block, consisting mainly of pieces from the 9-cube, and a $(7 \times 13 \times 10)$-block, consisting of pieces from the 10-cube. The 10-cube contributes piece C to the former block and has pieces D and E rearranged via a steplike approach to give the latter block. The relative positions of pieces G, H, and I do not change, and similarly for pieces L and K. To produce the 12-cube, Cadwell placed the 10-cube in one corner of the 12-cube and rotated the flat piece C by $90°$. This leaves three faces of the 12-cube to be accounted

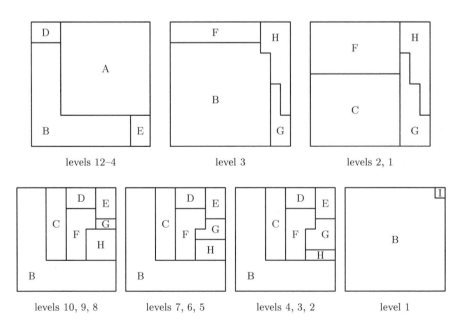

21.5: 12-cube and 10-cube in Cadwell's $12^3 + 1^3 = 10^3 + 9^3$

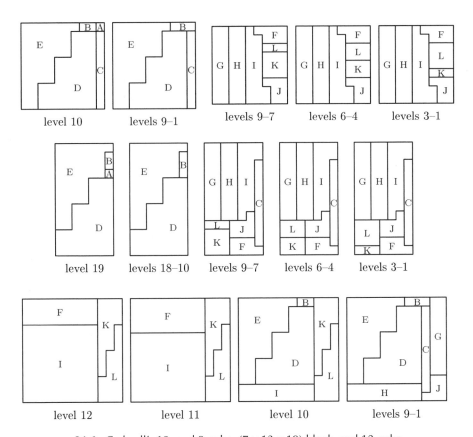

21.6: Cadwell's 10- and 9-cube, (7 × 13 × 19)-block, and 12-cube

for. Pieces G and J will clad most of one face, piece H clads most of another, and pieces F and I clad most of the third. This leaves a $3 \times 3 \times 12$ cavity, which pieces K and L will fill by using the step technique.

The relationship $1^3 + 1^3 + 5^3 + 6^3 = 7^3$ presents a good opportunity for a dissection, and Robert Reid obliged with an 8-piece one, which is clearly minimal for rational dissections. Reid left the two 1-cubes and the 5-cube uncut. The 7-cube and 6-cube are shown in Figure 21.7. Especially clever are the three-dimensional steplike cut involving pieces E and F and the convenient way that the two 1-cubes fit on the ends of piece F in the 7-cube.

Robert Reid found two other cube dissections that use a fairly small number of pieces. He has a 10-piece dissection of cubes for $3^3 + 10^3 + 18^3 = 19^3$ and an 11-piece dissection for $16^3 + 2^3 = 15^3 + 9^3$. Each of these uses two more pieces than the simple lower bounds discussed earlier. Will they stand impregnable or be beaten by another impatient generation?

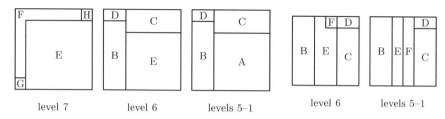

level 7   level 6   levels 5–1   level 6   levels 5–1

21.7: 7-cube and 6-cube in Reid's cubes for $1^3 + 1^3 + 5^3 + 6^3 = 7^3$

We now step away from numerical identities involving sums of cubes, and examine Thomas H. O'Beirne's three-dimensional application of the step dissection. O'Beirne (1961) described a set of six pieces that form six different rectangular blocks. These are a $(16 \times 18 \times 6)$-block, a $(16 \times 12 \times 9)$-block, a $(12 \times 24 \times 6)$-block, a $(12 \times 12 \times 12)$-block, a $(8 \times 24 \times 9)$-block, and a $(8 \times 18 \times 12)$-block, as shown in Figure 21.8. Each block is related to two of the other blocks and derives from

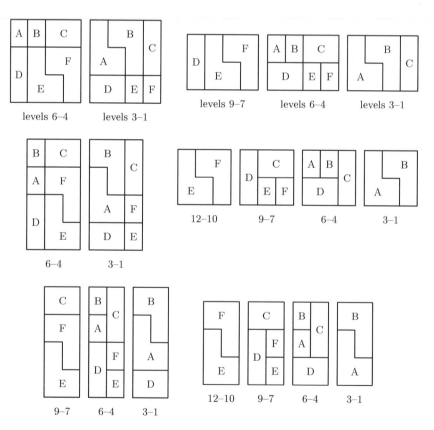

21.8: O'Beirne's six rectangular blocks

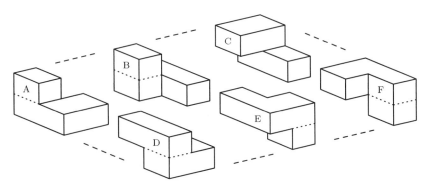

21.9: Perspective view of O'Beirne's six pieces

them by applying a step dissection. For example, the $(8 \times 18 \times 12)$-block in the lower right corner is produced from the cube above it by projecting a two-step cut through the third dimension. The block in the lower right corner also comes from the block in the lower left corner via a two-step dissection projected through the second dimension. In fact, the six blocks in Figure 21.8 are arranged in a cycle so that we can produce any block from either of its two neighbors in the cycle by a two-step dissection projected through the appropriate dimension. It is easy to see that these dissections are translational.

Pieces A through F are shown in perspective in Figure 21.9. We can slide them together as indicated by the dashed edges to give the block in the upper right corner of Figure 21.8. The pieces come in mirror image pairs: A and F, B and E, and C and D.

There is no reason to limit rational dissections of three-dimensional figures to cubes and rectangular solids. Our final dissection is of "layered figures" based on an identity that was noted by the nineteenth-century

Thomas Hay O'Beirne was born in Glasgow, Scotland, in 1915. He was awarded an M.A. in mathematics and physics by the University of Glasgow in 1938. For seven years he was a civilian employee of the Royal Naval Scientific Service, followed by three years as scientific officer for the Ordnance Survey of Great Britain, before becoming in 1949 chief mathematician for Barr & Stroud Ltd., a firm specializing in scientific instruments and precision engineering.

O'Beirne wrote a weekly series, "Puzzles and Paradoxes," for the *New Scientist* during 1961–1962. His book, based on this series and sharing the same title, was published in 1965. He presented some of his math puzzles on British and Dutch television and was interested in programming and recording computer music. O'Beirne viewed himself as a nationally conscious Scot.

French mathematician Georges Dostor: For each positive n,

$$(2n^2 + n)^2 + (2n^2 + n + 1)^2 + \cdots + (2n^2 + 2n)^2 = (2n^2 + 2n + 1)^2 + \cdots + (2n^2 + 3n)^2$$

This identity suggests an interesting three-dimensional dissection problem. For each square, create a rectangular solid of height 1 and then stack the rectangular solids, smaller on top of larger, with the upper left corner of the corresponding squares positioned directly above each other. We call the resulting figure a *stepped platform*, for obvious reasons. More specifically, an (a, b)-platform results when we use squares of sides $a, a + 1, \ldots, b$. We pose a natural problem: to dissect a $(2n^2 + n, 2n^2 + 2n)$-platform to a $(2n^2 + 2n + 1, 2n^2 + 3n)$-platform.

I have found a method for this problem that uses $n + 3$ pieces. We see the method in Figure 21.10 for the case $n = 2$, when $10^2 + 11^2 + 12^2 = 13^2 + 14^2$. We use round dots to indicate the corners that are directly over each other. Pieces A, B, and C extend through all levels of the $(2n^2 + 2n + 1, 2n^2 + 3n)$-platform. When moving pieces from one platform to the other, we flip pieces A and B over and shift piece B

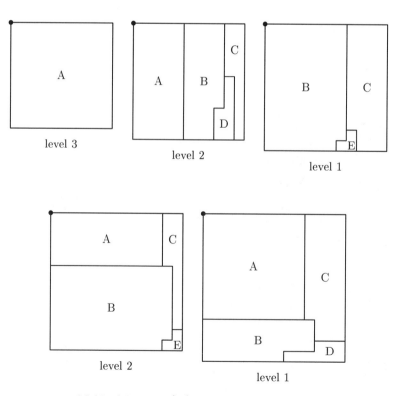

21.10: A $(10, 12)$-platform to a $(13, 14)$-platform

vertically one level with respect to A. Piece A has widths that are multiples of $2n + 1$ on its various levels, piece B has widths that are multiples of $2n$, and piece C has widths that are multiples of 2. In addition, there are n step-shaped pieces, where the ith piece has n steps of rise 1 and tread $2i - 1$. We flip each of these over in moving from one platform to the other. Perspective views of both platforms, with the boundaries between pieces indicated by bold lines, are shown in Figure 21.11.

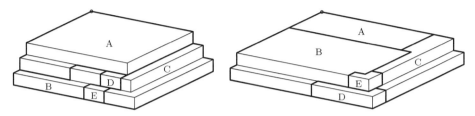

21.11: Perspective view of $(10, 12)$-platform to $(13, 14)$-platform

CHAPTER 22

Prisms Reformed

*It was a long commute that summer of 1973: onto the
Baltimore Beltway, then Interstate 95, then the Washington Beltway, then into Dupont Circle, and finally the tedious search for free on-street parking. And after 5:00 P.M.
I retraced the route in the reverse direction, every day,
five days a week. At work, I scrutinized the blueprints
of a Long Island hospital to unravel the layout of de-
partments (or were they archduchies?) spread through a
maze of interconnected buildings. Sometimes a department extended up through several levels of one building,
sometimes it extended across one level in several buildings, and sometimes both. At home, relaxation was af-
forded by a curious little book of geometrical drawings.
And one day, as the symmetries in the dissection of a
10-pointed star and two 5-pointed stars were reflected
and refracted in a mind game of what-if, the stars first
replicated, then rotated, then stacked as levels of a building, then merged into volumes worthy of those medical
potentates.*

In this chapter we explore what we can do when planar figures are given thickness,
that is, they are turned into right prisms (see Chapter 2). This happens whenever
we make a physical model of a planar dissection. We exploit this situation by making cuts parallel to the plane of thickness, as well as perpendicular to it. Of course,
our goal is to reduce the number of pieces from the corresponding planar dissection. The particular prism dissection that I discovered on that summer day in the
springtime of my life is not the simplest example of the dissections in this chapter.
Therefore we will study some other examples before we get to it in Figure 22.8.

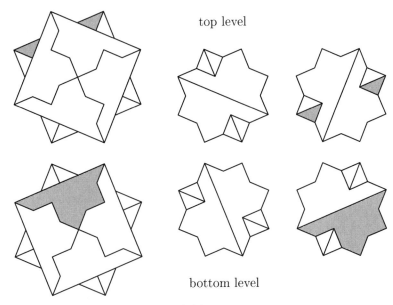

top level

bottom level

22.1: Two {8/2}-prisms to one

All dissections in this chapter except the one in Figures 22.3 and 22.4 originally appeared in (Frederickson 1979).

What planar dissections can benefit by adding thickness? The dissection of two {8/2}s to one in Figure 17.2 uses 11 pieces. By adding the third dimension, we can save three pieces and introduce a lot of symmetry. My dissection of two {8/2}-prisms to one is given in Figure 22.1. It is a two-level elaboration of the dissection of two equal squares to one that was shown in Figure 4.3. On each level, we slice the small {8/2}s in half, as we did with the squares in Figure 4.3. This alone is not enough, because the points in the small {8/2} get in the way of each other when we try to assemble them, and four of the points are missing in the large {8/2}. These problems disappear if we cut half-square pieces out of the small {8/2}s to make them fit together and use the half-square pieces to form the other four points in the large {8/2}. We can glue each half-square piece onto a main piece on the other level, forming the upper part of a "tower." Each of the resulting eight pieces has a central "bridge" that is on one level and that spans between two towers. The towers extend through the full thickness of the prisms and have half-square cross sections.

The dissection has 4-fold rotational symmetry in the large {8/2}-prism, and 2-fold rotational symmetry and 2-fold replication symmetry in the small {8/2}-prisms. It also has translation with no rotation and top versus bottom symmetry. A dissection has *top-versus-bottom* symmetry when it is identical with a copy of itself that

259

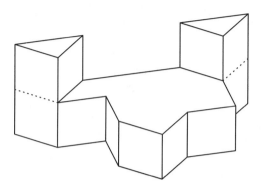

22.2: Perspective view of piece for two {8/2}-prisms to one

has been flipped to rest on its top and then rotated appropriately. All eight pieces are identical. A piece is shown shaded in Figure 22.1, and in perspective in Figure 22.2.

The dissection of two {12/3}s to one in Figure 17.6 is also based on the dissection of two squares to one. The third dimension helps in this dissection, too, saving an additional four pieces, as shown in Figure 22.3. The approach for the {12/3}-prisms is similar to that for the {8/2}-prisms, but the cuts across the small {12/3}-prisms are not simple straight slices, and there are now towers of three different shapes.

The dissection of {12/3}-prisms has the same nice symmetries as the dissection of {8/2}-prisms. Once again, all eight pieces are identical. A piece is shown shaded in Figure 22.3 and in perspective in Figure 22.4.

Can the the third dimension also help a three-to-one dissection? Yes, as shown by the 13-piece dissection of three {12/2}-prisms to one in Figure 22.5. It reduces by five the number of pieces used by a planar dissection such as the one in Figure 17.12.

The dissection is a two-level elaboration of a 7-piece (nonoptimal) dissection of three hexagons to one. That dissection slices two of the small {12/2}s in thirds and arranges the pieces around the third hexagon. Treating the {12/2} as a hexagon does not work, because the points in the small {12/2} get in the way of each other when we try to assemble them, and six of the points are missing on the large {12/2}. These problems disappear if we cut pieces out of the small {12/2}s that are 60°-rhombuses and halves of these rhombuses, and use these to form the towers. The corresponding internal structure, with irregular octagons not needing to be further subdivided, is shown in Figure 22.6.

The dissection has 6-fold rotational symmetry in the large {12/2}-prism, and 3-fold rotational and 2-fold replication symmetry for two of the small {12/2}-prisms.

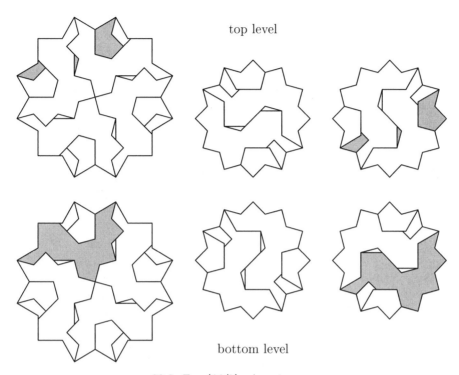

top level

bottom level

22.3: Two {12/3}-prisms to one

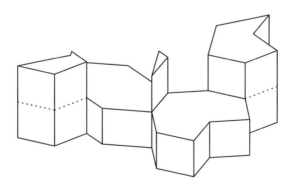

22.4: Perspective view of piece for two {12/3}-prisms to one

It also has top versus bottom symmetry. All pieces except the uncut {12/2}-prism are identical, and a perspective rendering of one of them is given in Figure 22.7. This piece is shown shaded in Figure 22.5.

Now that we have seen several examples of prism dissections, we take up the dissection (Figure 22.8) mentioned at the beginning of the chapter. There are six

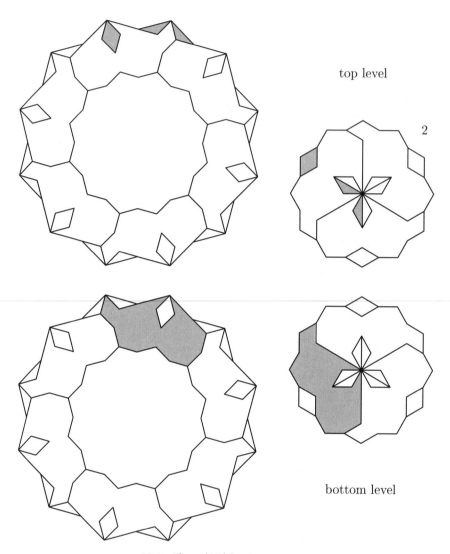

top level

2

bottom level

22.5: Three {12/2}-prisms to one

levels, with the top three shown in Figure 22.8 and the bottom three levels (3b, 4, and 5) being mirror images of levels 3a, 2, and 1, respectively. Each level represents an 18-piece dissection of a {10/4} to two {5/2}s. Levels 2, 3a, 3b, and 4 are rotated copies of Lindgren's dissection of a {10/4} to two {5/2}s, and levels 1 and 5 are rotations of a small variation of the same. When we attach pieces to form the towers, the resulting number of pieces plummets to ten. One of the ten pieces is shown shaded in the top three levels.

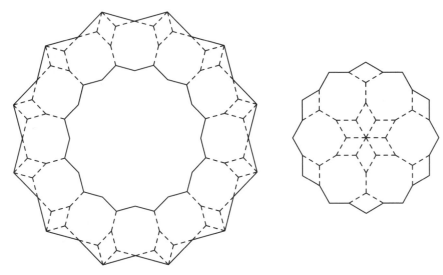

22.6: Internal structures of {12/2}s

The {10/4}-prism dissection is clearer if we compare Lindgren's dissection of a {10/4} to two {5/2}s to a dissection of a {10/3} to two {10}s. We show the latter dissection, using solid lines, in Figure 22.9. We extend the {10/3} to a {10/4} via the dashed edges, corresponding to adding twenty wedge-shaped pieces. Interestingly, we can extend the {10}s to {5/2}s by using the same twenty wedge-shaped pieces. Attaching the wedge-shaped pieces to the pieces of the {10/3} and the {10}s gives the 18-piece dissection represented by level 3a in Figure 22.8.

We attach the pieces on different levels by using the profiles of the wedge-shaped pieces. Thus each of the resulting ten pieces has a central "bridge" that is

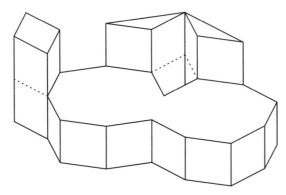

22.7: Perspective view of piece for three {12/2}-prisms to one

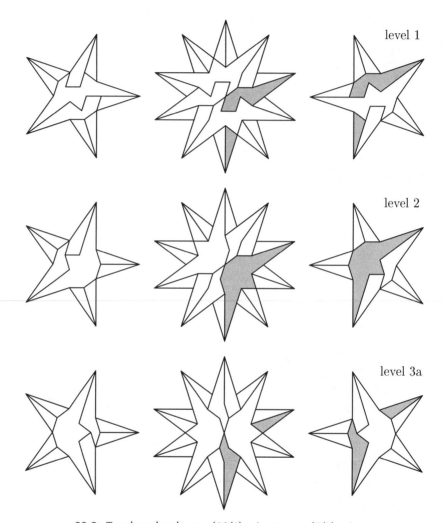

22.8: Top three levels: one {10/4}-prism to two {5/2}-prisms

on one level (counting 3a and 3b together as one level) and that spans between two "towers." The towers extend through the full thickness of the prisms and have wedge-shaped cross sections.

Why don't five copies of level 3a, suitably rotated, suffice to give the prism dissection? Although the pieces produced fill out the {10/4}-prism and the two {5/2}-prisms, they cannot be disentangled in three dimensions to shift from the {10/4}-prism to the {5/2}-prisms. This is the reason that levels 1 and 5 are slightly different from the others. This difference allows the pieces to be lifted vertically out of the {10/4}-prism and then dropped vertically into the two {5/2}-prisms.

22.9: {10/4} to two {5/2}s, from {10/3} to two {10}s

We can move the pieces from the {10/4}-prism to the two {5/2}-prisms without rotation. Furthermore, since the bottom three levels are mirror images of the top three, the dissection has top-versus-bottom symmetry. Since each level in the {10/4}-prism has 2-fold rotational symmetry, and each level in the two {5/2}-prisms has 2-fold replication symmetry, so do the respective prisms themselves. The four pieces with a bridge on level 1 or 5 are identical, the four pieces with a bridge on level 2 or 4 are identical, and the remaining two pieces are identical. A piece with a bridge on level 2 is shown in Figure 22.10, with the levels indicated by dotted lines. This is the shaded piece in Figure 22.8.

Another way to take advantage of the third dimension is to start with a dissection of $p_1 > 1$ copies of figure F_1 to $p_2 > 1$ copies of figure F_2 and then stack the like copies. Taking $p_1 = 3$ and $p_2 = 2$, we work on a prism P_1 that is 3 units high and a prism P_2 that is 2 units high. Except in the most unusual of cases, we would expect to save some pieces, so the goal should be to find a dissection in which the

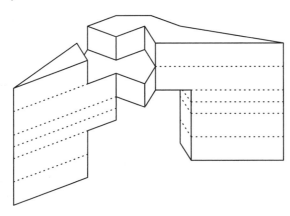

22.10: View of a piece for {10/4}-prism to two {5/2}-prisms

265

22.11: 2-high {12}-prism to 3-high {12}-prism

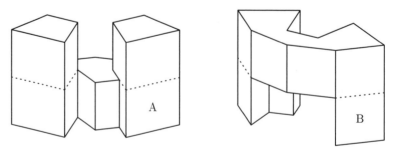

22.12: Perspective of pieces A, B for 3-high {12}-prism to 2-high

saving is substantial. This happens with dodecagons, as shown in Figure 22.11. It has seven fewer pieces than the planar dissection in Figure 17.26.

We base the dissection on a 23-piece (nonoptimal) dissection of three dodecagons to two given by Lindgren (1964b). In that dissection, Lindgren cut two of the three small dodecagons into four pieces each, with three of the four pieces being bigger and fitting on the boundaries of the larger dodecagons. He cut the remaining small dodecagon into fifteen pieces that filled out the larger dodecagons. Twelve of these fifteen pieces form towers for the six bigger pieces. We cut cavities in the bigger pieces to accommodate the towers from the other level. Perspective views of pieces A and B from Figure 22.11 are shown in Figure 22.12.

Greg Norman Frederickson was born in Baltimore, Maryland, in 1947. After graduating with an A.B. in economics from Harvard University, he taught mathematics in the Baltimore public schools for three years before entering graduate school. He received a Ph.D. in computer science from the University of Maryland in 1977. Since then he has been a faculty member in the computer science department first at Pennsylvania State University, and then at Purdue University. A full professor since 1986, he teaches and does research in the area of the design and analysis of algorithms and, in particular, graph algorithms and data structures.

Originally a tennis player, Frederickson now plays squash in a manner to be expected of an enthusiastic convert. He has dabbled in interior design, including a pentagonal coffee table that displays a bronze casting of a three-dimensional surface based on Penrose tiles. (These projects sometimes seem to get out of hand, as readers of this present volume may attest.) Much to the delight of his two young children and wife, he keeps a large raspberry patch, as well as some blueberry bushes and cherry trees.

CHAPTER 23

Cheated, Bamboozled, and Hornswoggled

The smart alec rudely interrupts the speaker, ready to spoil the puzzle for everyone else. But he has got the puzzle wrong and does not know it. And he is squelched by a cool response. Thus did Sam Loyd frame the anecdote accompanying his mitre-to-square dissection puzzle. And Loyd delighted in quoting the Persian proverb "He who knows not, and knows not that he knows not, is a nuisance."

But, ironically, it was Loyd, the audacious American who had tormented the world with the notorious 14–15 puzzle and his perplexing Get Off the Earth paradox, who had got it wrong. And it was the cool Englishman, Henry Dudeney, who took his pleasure in squelching Loyd's bogus 4-piece solution.

What was perhaps Sam Loyd's biggest goof occurred with the Smart Alec Puzzle. Loyd (*Inquirer*, 1901d) posed the puzzle, accompanied by a picture of a rude young man who interrupts the speaker to show off his knowledge. The attention-getting premise was that the smart aleck had assumed the puzzle to be the familiar one of cutting a mitre into four pieces of identical shape, whereas it was actually a new puzzle, that of dissecting a mitre into the fewest number of pieces that rearrange

THE ALEC SMART PUZZLE

BY SAM LOYD

PROPOSITION—Cut the mitre-shaped piece of paper into the fewest possible number of pieces which will fit together and form a perfect square.

into a square. The artwork that accompanied the puzzle statement appears on the previous page. It has a charm that is irresistible.

Loyd's purported 4-piece solution, shown in Figure 23.1, was given in (Loyd, *Inquirer*, 1901e). And here the trouble begins, because Loyd's solution does not yield a square. Loyd cut the mitre, which is exactly three-fourths of a square, as follows: First, he cut the top half off of each point, and rotated the two resulting triangles to fill the remaining part of the notch, giving a rectangle whose length is 4/3 its width. Then he used the step technique to convert the rectangle into what appears to be a square. But the resulting rectangle has a width that is actually 49/48 times its length!

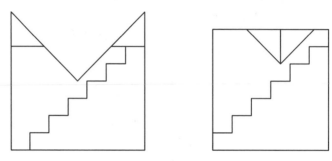

23.1: Loyd's mitre to purported square

Henry Dudeney (*Strand*, 1911a) pointed out Loyd's gaffe and gave the 5-piece dissection of Figure 23.2 in (Dudeney, *Strand*, 1911b). We can view Dudeney's solution as just an elaboration of Kelland's gnomon to square dissection in Figure 4.12. Dudeney first clipped off the left point and rotated it to fit flush against the right point, giving Kelland's gnomon. Then he could apply Kelland's dissection.

Besides this mitre miscue, we have seen Loyd's crossed-up solution for the Cross and Crescent Puzzle in Chapter 15. Also, we have questioned the authorship

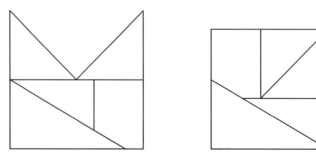

23.2: Dudeney's mitre to square

of the solution to his An Old Saw with New Teeth Puzzle in Chapter 15. So perhaps we should apply the same logic in the reverse direction here. Did Loyd's readers send him purported solutions to the Smart Alec and the Cross and Crescent puzzles that had fewer pieces than anticipated? And did he then, under deadline pressure, include them in his column without checking them carefully? The purported solution to the Smart Alec Puzzle in (Loyd, *Inquirer*, 1901e) had each of the four pieces numbered, and then a spurious number 5 sitting in the notch, serving no apparent purpose. Was the 5 a vestige of a 5-piece solution that Loyd had originally intended to publish?

Although unintentional, Loyd's erroneous solution is nonetheless a fraud on puzzledom. Consonant with the chapter title, we identify three types of fraud. The first, as we have seen, arises when an incorrect dissection is published in the belief that it is actually accurate. The serious puzzlist can rightly feel *cheated* when the purported solution is anything but.

Bamboozlement results from the second type of fraudulent dissection. This is a dissection that we present in the form of a paradox. We dissect a figure and rearrange the pieces to form another figure of supposedly larger or smaller area. Of course, the apparent gain or loss of area is offset by a complementary loss or gain elsewhere, which because of its shape risks being overlooked in a quick inspection.

The third type is one that appears to achieve fewer pieces by deliberately cheating ever so slightly in the dimensions of the actual figures. The reader is alerted to this fraud and invited to enjoy its artistry. Harry Lindgren used the term "approximate dissections," but I hope that he would have appreciated the appellation *hornswoggling*, which we apply to these dissections.

Having already been cheated, we progress to being bamboozled. William Hooper (1774) presented a paradox in which he purportedly converted a (10×3)-rectangle into two rectangles, of dimensions 4×5 and 6×2. Hooper called this creative exercise "geometric money." He cut the (10×3)-rectangle along one diagonal, then cut each resulting piece vertically, as shown in Figure 23.3. He then assembled like pieces, purportedly to give the desired rectangles. But each rectangle is "short" along a middle diagonal, as shown in Figure 23.4. Of course, when presenting the paradox, we should make it appear that the pieces fit to give the desired rectangles, either by making the diagonal lines in Figure 23.4 thick enough to cover up the shortfall or by drawing them incorrectly. However, with an increase in area of 2 over 30, or 6.7%, it is difficult to be convincing with this paradox.

Sebastiano Serlio (1551) hinted at a more convincing paradox. This also starts with a (10×3)-rectangle but takes it into a (7×4)-rectangle and a (3×1)-rectangle. He cut the (10×3)-rectangle as before on its diagonal, but the two remaining cuts are

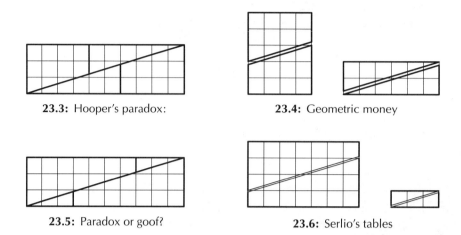

23.3: Hooper's paradox: **23.4:** Geometric money

23.5: Paradox or goof? **23.6:** Serlio's tables

different, as shown in Figure 23.5. The two resulting figures, drawn to emphasize the shortfall, are shown in Figure 23.6. The illusion is more effective than Hooper's, as the increase in area is 1 over 30, or 3.3%. Serlio was illustrating how he would create a wider table top. It appears that he was either unaware or unconcerned that the large rectangle was not a full 4 units wide. He did not explicitly mate the two small triangles.

Puzzle 23.1: Find a symmetrical 3-piece dissection of a (10×3)-rectangle into a (7×4)-rectangle and a (2×1)-rectangle.

Pietro Cataneo (1567) identified the error in Serlio's construction and pointed out the paradox of an increase in area. Cataneo, a Sienese architect and mathematician, also presented a variation of the problem for which Serlio's construction is correct. He dissected a table top that is a (12×4)-rectangle to a (9×5)-rectangle and a (3×1)-rectangle, using Serlio's approach. Cataneo's variation also admits a 3-piece dissection, similar to the solution to Problem 23.1, but not symmetrical.

The next paradox is more finely crafted, and consequently more convincing at first glance. Schlömilch (1868) dissected an 8-square into four pieces that he rearranged to form, purportedly, a (13×5)-rectangle. Loyd (*Eagle*, 1897a) mentions this paradox, relating it to a chessboard. Loyd (1914, p. 288) claimed that he presented it before the first American Chess Congress in 1858. Although Fiske (1859) identified Loyd as receiving honorable mention (i.e., third place) in the problem tournay and also being a subscriber of the Congress, he made no mention of Loyd's having presented the chessboard paradox. The dissection of the 8-square is shown in Figure 23.7, and the pieces are shown rearranged in Figure 23.8. When the paradox is illustrated pictorially, the open area in the middle of the (13×5)-rectangle is replaced by a diagonal. Since the gain in area is only 1 over 64, or 1.6%, the paradox is relatively convincing.

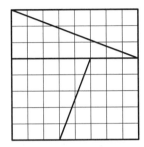

23.7: Chessboard:

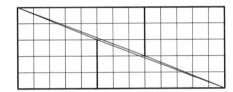

23.8: Expansion paradox

This paradox is based on the relationship of the Fibonacci numbers (1, 2, 3, 5, 8, 13, 21,...) to powers of the golden ratio $\phi = (1 + \sqrt{5})/2$. Given the first two Fibonacci numbers, each succeeding one is the sum of the previous two. As for powers of ϕ, it follows from the definition of ϕ that $\phi^i = \phi^{i-1} + \phi^{i-2}$. The dissection of an 8-square includes two rectangles, of dimensions 3×8 and 5×8. If we were to cut it into an $((8/\phi^2) \times 8)$-rectangle and an $((8/\phi) \times 8)$-rectangle, we could rearrange the pieces to give exactly an $((8/\phi) \times (8\phi))$-rectangle. In this case, we have "created" no area!

Walter Dexter (1901) gave an interesting variation of the paradox. He dissected the 8-square as in Figure 23.7 but rearranged the pieces purportedly to form a figure of area 63, rather than 65. In the rearrangement in Figure 23.9 the pieces overlap ever so slightly. Also, two of the pieces project slightly over what should be right-angled cavities. Discounting the overlap and these projections does indeed give an area of 63. The region of

273

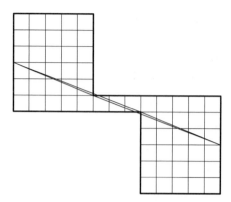

23.9: Chessboard: at a loss

overlap and projection is of precisely the same shape as the missing area in the (13×5)-rectangle. As before, the paradox relies on drawing a line that obscures the fact that the pieces overlap. Again, if we were to use the golden ratio when cutting the 8-square, then no overlap would result, and the paradox would vanish.

Since we have been bamboozled with two-dimensional dissection paradoxes, what about three dimensions? James Propp, a mathematics professor at MIT, asked whether there was an interesting paradox based on the fact that $5^3 + 6^3$ is just 2 smaller than 7^3. David Singmaster proposed $6^3 + 8^3$ versus 9^3, but I have settled on $9^3 + 10^3$, which is just 1 larger than 12^3. Even though James Cadwell gave an exact dissection for $9^3 + 10^3 = 12^3 + 1^3$, why should this discourage us from attempting a "proof" that $9^3 + 10^3 = 12^3$? Maybe it will dampen all that brouhaha about the recent proof of Fermat's famous "last theorem," that $x^n + y^n = z^n$ does not have an integral solution for any integer $n > 2$.

My dissection paradox (also discovered independently by Anton Hanegraaf) is shown in Figure 23.10, with the 9-cube on the left, the 10-cube on the right, and the 12-almost-cube in the center. I have placed hashmarks at unit intervals along some of the edges to make distances easier to gauge. The fact that both $(12 - 9)$ and $(12 - 10)$ divide evenly into 12 allows us to use the step technique in this dissection. First convert the 9-cube by a step dissection to the block on the lower left, then by another step dissection into the block constituting the left side of the 12-almost-cube. Since these two steps intersect, four pieces (A, B, C, and D) result. Then convert the 10-cube by a step dissection to the block on the lower right, and then by another step dissection into the block constituting the right side of the 12-almost-cube. Again four pieces (E, F, G, and H) result. Whereas a dissection for $9^3 + 10^3 = 12^3 + 1^3$, such as James Cadwell's, must use nine pieces, this dissection paradox uses only eight pieces. But where is the paradox? The resulting block is

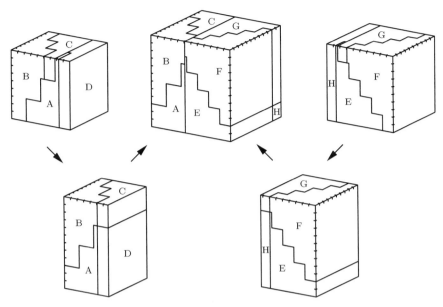

23.10: Fermat upended?: $9^3 + 10^3 = 12^3$?

$(1729/1728)12 \times 12 \times 12$, with the long dimension 1.0058 times longer than the others. Would Fermat have been bamboozled?

Having thoroughly bamboozled ourselves, we cut our intellectual losses and move on to being hornswoggled. Lindgren (1953) gave a number of approximate dissections and later (1968) focused exclusively on them. Perhaps the best known is a 5-piece dissection of a pentagon to a near-square. This dissection owes its infamy to comments by Dudeney (1917), originally in (*Dispatch*, 1903a), regarding exact dissections of a pentagon to a square:

> I have received what purported to be a solution in five pieces, but the method was based on the rather subtle fallacy that half the diagonal plus half the side of a pentagon equals the side of a square of the same area. I say subtle, because it is an extremely close approximation that will deceive the eye, and is quite difficult to prove inexact. I am not aware that attention has before been drawn to this curious approximation.

Lindgren (1953) concluded from Dudeney's discussion that the dissection was probably derived as follows: He partitioned the pentagon as on the left of Figure 23.11 into pieces that form the parallelogram in the center of the figure. Then he converted the parallelogram to a rectangle by cutting along the dashed line and converted the rectangle to what appears to be a square by a P-slide. The resulting

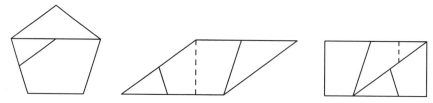

23.11: Pentagon to parallelogram to rectangle, then fake P-slide

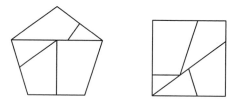

23.12: Pentagon to near-square

5-piece dissection is given in Figure 23.12. But the P-slide that converts the rect-angle to a true square does not coincide with the diagonal from the rectangle's upper right corner. In fact, the ratio of the width to the length of the near-square is $(10 - 4\sqrt{5})\sin 72° \approx 1.004$. Philip Tilson (1979) presented the same approximate dissection but derived it by using strip techniques, noting that two lines almost, but do not actually, coincide.

The pentagon is fertile ground for hornswoggling. The *AMM* problem editor pointed out in (Goldberg 1952) that W. B. Carver had found a 6-piece approximate dissection of a pentagon to an equilateral triangle. This approximation relies on the fact that the side length of the triangle is just slightly less than twice the side length of the pentagon. The ratio is $2\sqrt{\sin 72°(5 + \sqrt{5})}/\sqrt{3} \approx 1.9933$. A triangle resulting from using the approximate value of 2 would be an isosceles triangle that is almost equilateral. Lindgren (1953) took the pattern for the pentagon that appears in Figure 11.5 and crossposed its strip with the strip for the isosceles triangle. Without recourse to strips, we can convert a parallelogram to a triangle

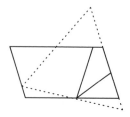

23.13: Partition {5}

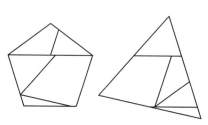

23.14: Pentagon to near-triangle

of the same height and specified angles. The technique is shown applied in Figure 23.13, with the right side of the parallelogram viewed as its base, and the altitude of the resulting triangle extending off to the left. The dotted lines indicate cuts of the parallelogram and the outline of the resulting triangle. Lindgren's 5-piece approximation appears in Figure 23.14.

CHAPTER 24

Solutions to All Our Problems

☞ *The young students are particularly recommended to give every question a fair trial, before they refer to the key for assistance. The same hint may be given to those of riper years, who also, it is hoped, will find some amusement in this little volume.*

This admonition appears at the end of John Jackson's preface to his (1821) book *Rational Amusement for Winter Evenings*. I heartily proffer this same advice, but having done so, proceed posthaste to the solutions.

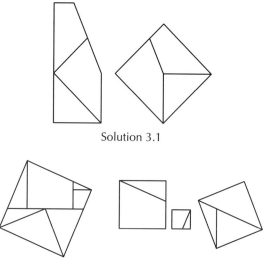

Solution 3.1

Solution 4.1

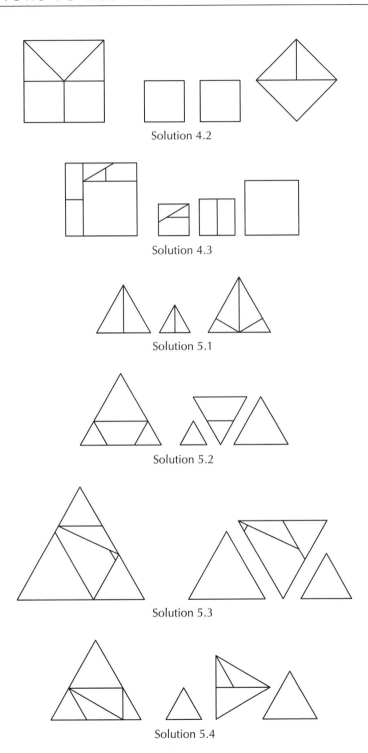

Solution 4.2

Solution 4.3

Solution 5.1

Solution 5.2

Solution 5.3

Solution 5.4

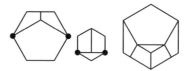

Solution 5.5. (Hinges are shown in the two smaller hexagons.)

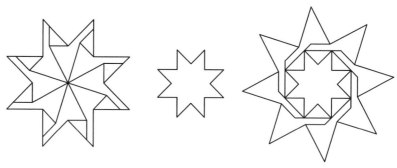

Solution 5.6

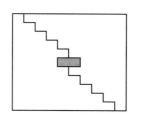
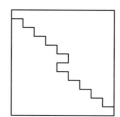

Solution 6.1

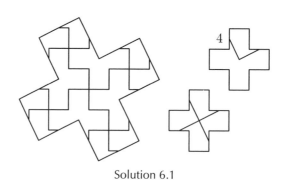

Solution 7.1. By Dudeney (*Strand*, 1926c).

Solution 7.2. This generalizes the dissection in Figure 7.6 to all squares in the class of Pythagoras:

1. Cut an x-square out of the upper left corner of the z-square.
2. Cut a $(z - x \times x)$-rectangle from the lower left of the z-square.

3. Stair-step starting 1 below the upper right corner of the z-square:
   Move/cut to the left a distance of 2n.
   [n − 1 times]: {Move/cut down 2m; Move/cut left 2n.}

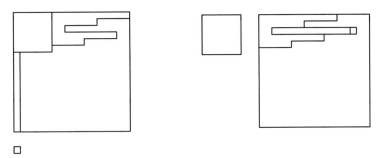

Solution 8.1. (Modified from Figure 8.4.)

Solution 8.2. When $2p - q = 0$, the method for the Penta class yields a dissection for $3^2 + 4^2 = 5^2$ with the 4-square left intact.

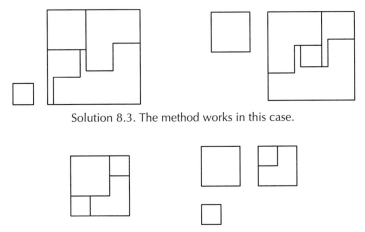

Solution 8.3. The method works in this case.

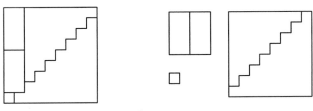

Solution 8.4. Method 6 applies, with $n = 1$. The method produces a rectangle of height 0, which doesn't count as a piece. See (Dudeney 1907).

Solution 8.5

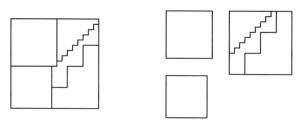

Solution 8.6. Other solutions exist.

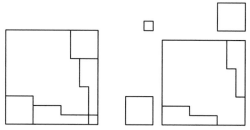

Solution 8.7. Example for $n = 3$. For general n, step pieces have $n-1$ steps of length n and height 1. A different method for $n = 2$ uses only five pieces.

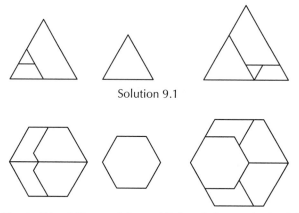

Solution 9.1

Solution 9.2. Due to Friend Kierstead, Jr., and Robert L. Patton, Jr., in (Schmerl 1973).

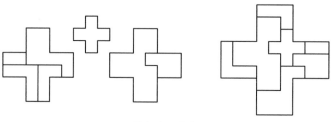

Solution 9.3

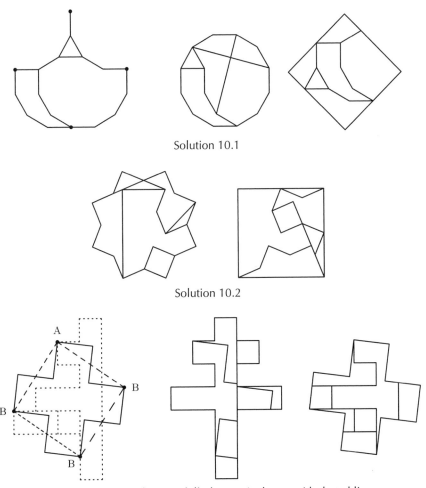

Solution 10.1

Solution 10.2

Solution 10.3. For reference, {L'} element is shown with dotted lines.

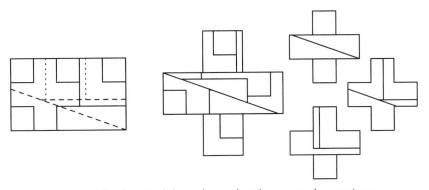

Solution 11.1. Use a P-slide as shown, but then reattach two pieces.

283

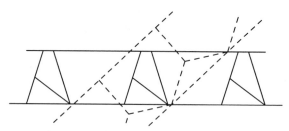

Solution 11.2. Lindgren (1964b) gave the corresponding 7-piece dissection.

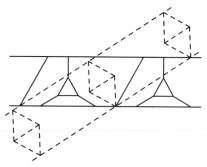

Solution 11.3. My 9-piece dissecton is in the 1991 revision of (Gardner 1969).

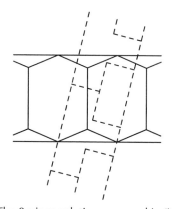

Solution 11.4. The 8-piece solution appeared in (Lindgren 1951).

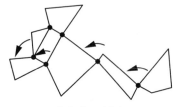

Solution 12.1

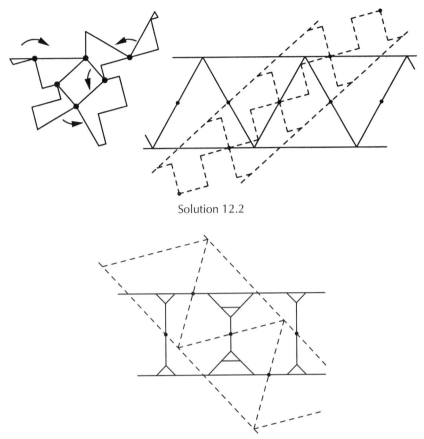

Solution 12.2

Solution 12.3. Crossposition of strips for {8} and {3}.

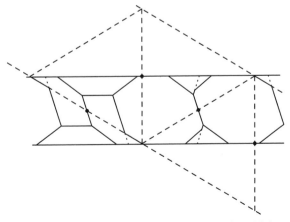

Solution 12.4. Crossposition of strips for {10} and {3}.

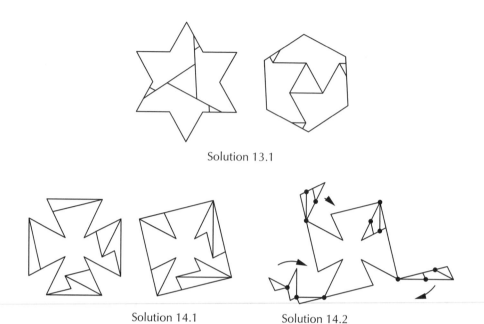

Solution 13.1

Solution 14.1   Solution 14.2

In Solution 14.2 the thinnest small piece is obstructed by the arm of the cross. Use a hinge that has some "play" in it, so that the small piece may be lifted up into the third dimension, rotated into the correct orientation, and then dropped down into final position.

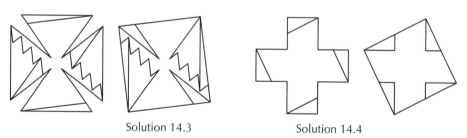

Solution 14.3   Solution 14.4

In Solution 14.4, the cross on the 3 x 3 grid corresponds to {G}. (See (Dudeney 1917).)

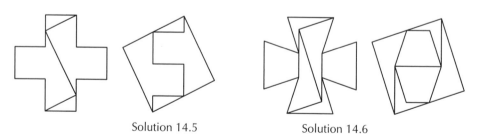

Solution 14.5   Solution 14.6

In Solution 14.6, the dissection for the 4 x 4 grid is due to Bernard Lemaire and appears in (Berloquin, *Le Monde*, 1975b).

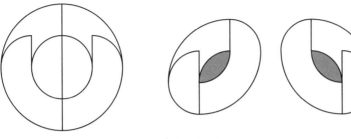

Solution 15.1

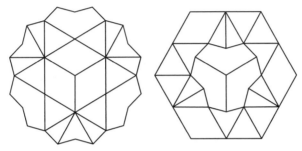

Solution 16.1. (By using some of the tricks in Figure 16.6, the number of pieces can be reduced from 21 to 15.)

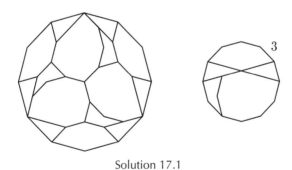

Solution 17.1

Solution 17.2. Hinged, too!

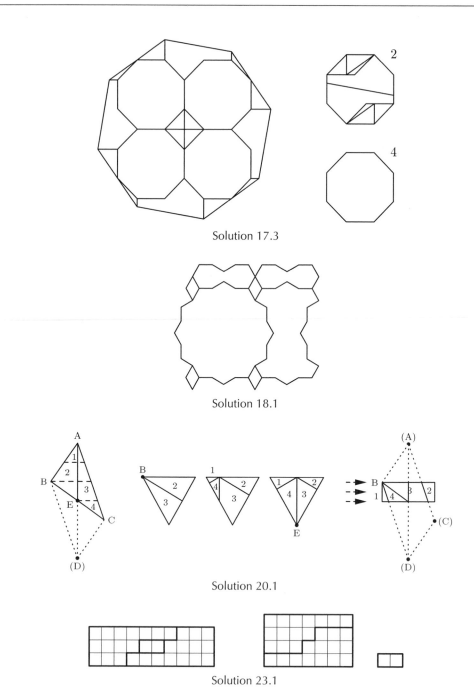

Solution 17.3

Solution 18.1

Solution 20.1

Solution 23.1

Afterword

Needham (1959) gave additional information about the Chinese dissection that estimates π. Anonymous (1897) and Newing (1994) provided evidence of the strain in the relationship of Loyd and Dudeney. The graphics and/or text of Loyd's Sedan Chair Puzzle, Guido Mosaics Puzzle, Mrs. Pythagoras' Puzzle, the Cross and Crescent, the Puzzle of the Red Spade, and The Smart Alec Puzzle are reproduced from (Loyd 1914).

An English translation and commentary for Kepler's polygon tilings appeared in (Field 1979). A biography of Kepler's life with a careful explication of his work can be found in (Caspar 1959), and a survey of Kepler's mathematical contributions is given in (Coxeter 1975). Hogendijk (1984) surveyed early knowledge of the structure of polygons with an odd number of sides. A comprehensive treatment of tessellations and tilings can be found in (Grünbaum and Shephard 1987). See Lines (1935), Coxeter (1963), and Wenninger (1971) for discussions of semiregular polyhedrons. Kepler (1940) first discussed the rhombic dodecahedron, which is also mentioned in (Coxeter 1963).

I rolled out some heavy machinery at the beginning of Chapter 3. Perhaps I should belatedly include a warning by Sir Geoffrey Keynes in (Blake 1967) about interpreting "The Tyger": "It seems better, therefore, to let the poem speak for itself, the hammer strokes of the craftsman conveying to each mind some part of his meaning.... Careful dissection will only spoil its impact as poetry."

The title for Chapter 4 is the title of a song from the children's television show *Sesame Street*. Sayili (1958) gave the Arabic text, Turkish translation, explanation, and analysis of Thābit's dissection and he also gave (1960) a discussion in English. Böttcher (1921) surveyed the history surrounding Thābit's dissection and suggested that the dissection might be of Indian origin. Al-Nayrizi attributed the dissection to Thābit in his commentary on Euclid, appearing in translation in (al Nayrizi 1899). Perigal is remembered in (Anonymous 1899) and (Perigal 1901). Using Percy MacMahon's theory of repeating patterns, Macaulay (1922) gave other demonstrations of superposing tessellations. Cantor (1907, p. 185) and Allman (1889) discussed the dissection of two equal squares to one as it appears in Plato's works. Woepcke (1855) and Fourrey (1907) described Abū'l-Wafā's dissections of squares. Allman (1889) discussed the dissection of three equal triangles to one as it appeared in Plato.

An obituary for Harry C. Bradley appeared in (Anonymous 1936).

Ore (1953) wrote an engaging biography of Girolamo Cardano. Dickson (1920, pp. 165–166) gave a more complete discussion of the history of integer solutions for $x^2 + y^2 = z^2$. See pages 225 and 226 for the history of solutions for $x^2 + y^2 = z^2 + w^2$, and pages 261 and 265 for the history of solutions for $x^2 + y^2 + z^2 = w^2$. Kaufman-Rosen (1995) and Sager (1995) reported the story behind the invention of IBM's dissected keyboard. Loyd seems to have framed his puzzle on squares for $2^2 + 3^2 + 6^2 = 7^2$ too late for inclusion in the issues of *Our Puzzle Magazine*, which Sam Loyd, Jr., collected in *Cyclopedia of Puzzles* after his father's death in 1911. Dudeney's life and philosophy were discussed in (Sherie 1926), (Dudeney 1926b), and (Newing 1994). Bain (1907), Eaton (1911), and White (1913) discussed Loyd's life.

The dissection in Figure 10.2 was produced as a set of steel pieces and sold for 10 cents by the Oaks Magical Company in 1913. A photograph of the set is given in (Slocum and Botermans 1994). An obituary for Mott-Smith is given in (Anonymous 1960).

Gailliard (1985) gave a synopsis of Paul Busschop's life. Keynes (1937) wrote an obituary for Macaulay.

The presentation of the triangle-to-square dissection in (Dudeney, *London*, 1903a), and subsequently in (Dudeney 1907), fits in with the questions I raise about the dissection's origin. Dudeney had the haberdasher pose this puzzle to the pilgrims without knowing its solution. The haberdasher "narrowly escaped a sound beating," which could symbolize the puzzle columnist's plight if his correspondents discover that his own solution is less than the best. Did the fictional situation mirror actual experience? Was this a "Dudeneyan slip"? Lamb (1928) wrote an obituary for Taylor. Roberts (1996) wrote a fascinating account of Albert Harry Wheeler's interaction with mathematicians during the first half of the twentieth century.

An obituary for Freese appeared in (Anonymous 1957).

Portions of Chapter 14 are reproduced from Frederickson (1996–1997). Even though he wrote a book and two journal articles, P. Gerwien is somewhat of a mystery. Did he live a long life or a short one, and what did the P stand for? Also, who was A. E. Hill? Was he Alfred Ebsworth Hill (1862–1940), a London violin maker, who with his brothers wrote books about Antonio Stradivari and the violin makers of the Guarneri family? Or was he Arthur Edward Hill, an American scientist? Or someone else?

Judy Lindgren (1992) wrote an obituary for her father. I have rewritten the trigonometric relationships from Chapter 20 of (Lindgren 1964b) and (Frederickson 1972a) so that they hold for every positive integral value of n. For relationship (16.9) when $n = 2$, I gave in (Frederickson 1972a) a 14-piece dissection for {14/5} to {14/2}. For relationship (16.11) I gave in (1972a) an 18-piece dissection for two {9/2}s to one {18/7}. In (1972a) I gave an 18-piece dissection for two {7/3}s to one {14/2}. I have also found an 18-piece dissection of {14/5} to {7}. But I wonder whether one can do better, and thus do not reproduce it here. For an introduction to surface topology, see (Firby and Gardiner 1982) or (Gross and Tucker 1987).

Raymond Chandler was a renowned master of the detective genre, a keen observer of life in Southern California, and a contemporary of Ernest Irving Freese. His stylish novels included *The Big Sleep* and *Farewell, My Lovely*. Langford (1960) gave a different 8-piece dissection of two octagons to one; it is unclear how to generalize his approach to other {4n}s. Maxwell (1969) wrote an obituary for Langford. Lipton (1959) included a glossary of beatnik slang.

Grünbaum and Shephard (1987) discussed an expansion for Penrose tiles that is similar to Varsady's expansion of $\{3_{1/5}\}$ and $\{3_{2/5}\}$. Varsady (1986) explored relationships such as those between the number of pentagons and pentagrams on one level versus the number on the next level in his expansions. The square of the silvern ratio was called the "silver number" by Tutte; see (Le Lionnais 1983, pp. 51, 143). In analogy to ϕ and the Fibonacci numbers, ϑ is related to the numbers defined by the recurrence $G_0 = 0$, $G_1 = 1$, $G_2 = 2$, $G_n = G_{n-1} + 2G_{n-2} - G_{n-3}$.

A summary of Montucla's life appears in (Hutton 1840). An obituary for Wallace is given in (Anonymous 1844).

The exchange of letters between Gauss and Gerling is collected in (Gauss 1900). For a more complete account of the history and mathematics behind Hilbert's third problem, see (Boltyanskii 1978). In addition to his three classes of tetrahedra, M. J. M. Hill also described two isolated tetrahedra that are dissectable to a cube. Additional cases have been identified by Sydler (1956), Goldberg (1958), Lenhard (1962), Goldberg (1969), and Goldberg (1974). Cadwell (1966) illustrates the honeycomb formed from truncated octahedra.

Richmond was remembered by Milne and White (1949). See (Dickson 1920, p. 555) for the history of solutions for $x^3 + y^3 + z^3 = w^3$. Slocum and Botermans (1994) gave a photograph of a wooden model of O'Beirne's dissection that was constructed by Trevor Wood. See (Dickson 1920, p. 320) for a discussion of $(2n^2 + n)^2 + (2n^2 + n + 1)^2 + \cdots + (2n^2 + 2n)^2 = (2n^2 + 2n + 1)^2 + \cdots + (2n^2 + 3n)^2$.

The title for Chapter 23 is a quote from the cartoon show *The Adventures of Rocky and Bullwinkle*, at the beginning of the final episode of "The Weather Lady." Rocky summed up the heroes' dilemma: "We've been cheated, bamboozled, and hornswoggled." To which Bullwinkle responded, "Sounds like a law firm." Santaniello (1970) described Serlio's life. Slocum and Botermans (1987) gave a photograph of a wooden version of the chessboard paradox, produced around 1900 by AWGL of Paris, entitled "L'Échiquier Fantastique." Gardner (1956) discussed Hooper's paradox and the chessboard paradoxes, as well as a number of others. In presenting the paradoxes, Gardner made ample use of thick hand-drawn lines to obscure thin areas of overlap or thin areas left unfilled.

It was disappointing not to have located the final manuscript of Ernest Irving Freese. But perhaps it would have been anticlimactic, given the wonderful collections of dissections by Alfred Varsady, Anton Hanegraaf, Robert Reid, and Gavin Theobald. Time and space prevented including all of their work. I have resisted the impulse to produce a catalog of their unpublished dissections. The catalog itself would have been lengthy, and I would then have had to check the correctness of a great many constructions – interesting and fun, but not conducive to bringing this book to press expeditiously. Readers are left to trust my taste in selecting what I felt to be most suitable from these collections.

As I uncovered earlier sources for Dudeney and Loyd, I reattributed a fair number of dissections. Although the current attributions are to the earliest material that I know of, I would be interested but no longer surprised to learn of yet earlier references.

I plan to maintain a web-page about the book on the internet. The idea is to post new developments and items of interest. The URL will be

http://www.cs.purdue.edu/homes/gnf/book.html

I would like to hear of any additional historical or biographical information, and of any new or improved dissections. My network address is

gnf@cs.purdue.edu

Bibliography

al Nayrizi, Abu'l A. a.-F. i. H. (1899). *Anaritii In decem libros priores Elementorum Euclidis commentarii.* Euclidis Opera omnia. Supplementum. Lipsiae: B. G. Teubneri. Edited by Maximilian Curtze, from a translation from Arabic by Gherardo of Cremona. See p. 85.

Allman, George J. (1889). *Greek Geometry from Thales to Euclid.* Dublin: Hodges, Figgis & Co.

Anonymous (1844). Report of the council of the society to the twenty-fourth annual general meeting, held this day. *Monthly Notices of the Royal Astronomical Society* **6**, 25–44. Obituary for William Wallace on pages 31–41.

Anonymous (1897). Eagle puzzles abroad. *Brooklyn Daily Eagle* (February 7), p. 26. Adjacent to Loyd's puzzle column. Is it written by him?

Anonymous (1899). *Monthly Notices of the Royal Astronomical Society* **59**, 226–8. Obituary for Henry Perigal.

Anonymous (1936). Prof. H. C. Bradley dies after attack of heart. *The Tech.* (March 10).

Anonymous (1950). Two dissection problems. *Eureka: The Journal of the Archimedeans* (12), 6. Gardner (1986) stated that the problem was posed by John Leech.

Anonymous (1951). Solutions to problems in Eureka No. 13. *Eureka: The Journal of the Archimedeans* (13), 23. R. F. Wheeler's solution is to the second of the "Two Dissection Problems."

Anonymous (1957). Last tribute paid architect Ernest I. Freese. *Los Angeles Times* (October 23), Part 3, p. 3.

Anonymous (August 20, 1960). Geoffrey Mott-Smith dies at 58. *The New York Times*, p. 19.

Bain, George C. (1907). The prince of puzzle-makers. *The Strand Magazine* **34**, 771–7. About Sam Loyd.

Baker, H. F. (1944). Geoffrey Thomas Bennett. Obituary Notices of Fellows of the Royal Society.

Ball, W. W. R. (1939). *Mathematical Recreations and Essays* (11th ed.). London: Macmillan. Revised by H. S. M. Coxeter.

Berloquin, Pierre (*Le Monde*). En toute logique. Le Monde. Column in the fortnightly section Des Sciences et des Techniques. (1974a): March 27, p. 18; (1974b): April 10, p. 15; (1974c): May 8, p. 22; (1974d): May 22, p. 20; (1975a): Jan. 1, p. 8; (1975b): Jan. 15, p. 15; (1975c): April 23, p. 20.

Blake, William (1794). *Songs of Experience.* Originally published by W. Blake. The quoted text follows Copy Z, the Blake Trust facsimile. See *Blake's Poetry and Designs*, ed. by Mary Lynn Johnson and John E. Grant, W. W. Norton, 1979, New York.

Blake, William (1967). *Songs of Innocence and of Experience.* New York: Orion Press. Reproduction of Blake's illuminated book, with an introduction and commentary by Sir Geoffrey Keynes. See the commentary accompanying plate 42.

Boltyanskii, V. G. (1963). *Equivalent and Equidecomposable Figures.* Boston: D. C. Heath and Co. Translated and adapted from the first Russian edition (1956) by Alfred K. Henn and Charles E. Watts.

Boltyanskii, V. G. (1978). *Hilbert's Third Problem.* Washington, D.C.: V. H. Winston & Sons. Translated by Richard A. Silverman.

Boyai, Farkas (1832). *Tentamen juventutem.* Maros Vasarhelyini: Typis Collegii Refomatorum per Josephum et Simeonem Kali. In Latin.

Böttcher, J. E. (1921). Beweis des Tsabît für den pythagoreischen Lehrsatz. *Zeitschrift für mathematischen und naturwissenschaftlichen Unterricht* **52**, 153-60.

Bradley, H. C. (1921). Problem 2799. *American Mathematical Monthly* **28**, 186-7.

Bradley, H. C. (1930). Problem 3048. *American Mathematical Monthly* **37**, 158-9.

Bricard, R. (1896). Sur une question de géométrie relative aux polyèdres. *Nouvelles Annales de Mathématiques* **15**, 331-4, Series 3.

Brodie, B. (1884). Superposition. *Knowledge* **5**(135), 399.

Brodie, Robert (1891). Professor Kelland's problem on superposition. *Transactions of Royal Society of Edinburgh* **36**, Part II(12), 307-11 + plates 1 and 2. See Fig. 12, pl. 2.

Burdick, J. R. (1967). *The American Card Catalog.* Franklin Square, New York: Nostalgia Press.

Busschop, Paul (1876). Problèmes de géométrie. *Nouvelle Correspondance Mathématique* **2**, 83-4.

Cadwell, J. H. (1964). Some dissections involving sums of cubes. *Mathematical Gazette* **48**, 391-6.

Cadwell, J. H. (1966). *Topics in Recreational Mathematics.* Cambridge: Cambridge University Press.

Cadwell, J. H. (1970). A three-way dissection based on Ramanujan's number. *Mathematical Gazette* **54**, 385-7.

Cantor, Moritz (1907). *Vorlesungen über Geschichte der Mathematik* (3rd ed.), Volume 1. Stuttgart: B. G. Teubner. Reprinted 1965.

Cardano, Girolamo (1663). *Hieronymi Cardani Mediolanensis, philosophi ac medici celeberrimi,* Volume III: De Rerum Varietate. Lugdani, Sumptibus Ioannis Antonii Hugvetan & Marci Antonii Ravaud. Originally published in 1557. Other titles: Works. Opera omnia. See p. 248 (mislabeled p. 348).

Carpitella, Duilio (1996). Geometric Equidecomposition of Figures and Solids, manuscript.

Caspar, Max (1959). *Kepler.* London: Abelard-Schuman. Translated by C. Doris Hellman.

Catalan, Eugène (1873). *Géométrie Élémentaire* (5th ed.). Paris: Dunod. See p. 194.

Catalan, Eugène (1879). *Géométrie Élémentaire* (6th ed.). Paris: Dunod. See p. 237.

Cataneo, Pietro (1567). *L'Architettura di Pietro Cataneo Senese.* Venice. See pp. 164-5, in Libro Settimo.

Cheney, Wm. Fitch (1933). Problem E4. *American Mathematical Monthly* **40**, 113-14.

Cohn, M. J. (1975). Economical triangle-square dissection. *Geometriae Dedicata* **3**, 447-67.

Collison, David M. (1979-1980). Rational geometric dissections of convex polygons. *Journal of Recreational Mathematics* **12**(2), 95-103. Baywood Publishing.

Coxeter, H. S. M. (1963). *Regular Polytopes* (2nd ed.). New York: Macmillan.

Coxeter, H. S. M. (1975). Kepler and mathematics. In A. Beer and P. Beer (Eds.), *Kepler: Four Hundred Years,* Volume 18 of *Vistas in Astronomy,* pp. 661-70. Oxford: Pergamon Press.

Cundy, H. M. and C. D. Langford (1960). On the dissection of a regular polygon into n equal and similar parts. *Mathematical Gazette* **44**, 46.

Dehn, M. (1900). Über den Rauminhalt. *Nachrichten von der Gesellschaft der Wissenschaften zu Göttingen, Mathematisch-Physikalische Klasse,* 345-54. Subsequently in *Mathematische Annalen* **55**(1902) 465-78.

Dexter, Walter (1901). Some postcard puzzles. *The Boy's Own Paper* (December 14), 174–5.

Dickson, L. E. (1920). *History of the Theory of Numbers.* Volume II *Diophantine Analysis.* New York: Chelsea.

Dintzl, E. (1931). Über die Zerlegungsbeweise des verallgemeinerten pythagoreischen Lehrsatzes. *Zeitschrift für mathematischen und naturwissenschaftlichen Unterricht* **62**, 253–4.

Domoryad, A. P. (1964). *Mathematical Games and Pastimes.* New York: Macmillan Company. Translated from the original Russian, *Matmaticheskiye igry i razvlecheniya,* published in 1961 by Fizmatgiz, Moscow.

Dudeney, Henry E. (1907). *The Canterbury Puzzles and Other Curious Problems.* London: W. Heinemann. Revised edition printed by Dover Publications in 1958.

Dudeney, Henry E. (1917). *Amusements in Mathematics.* London: Thomas Nelson and Sons. Revised edition printed by Dover Publications, 1958.

Dudeney, Henry E. (1926a). *Modern Puzzles and How to Solve Them.* London: C. Arthur Pearson.

Dudeney, Henry E. (1926b). The psychology of puzzle crazes. *The Nineteenth Century* (December), 868–79.

Dudeney, Henry E. (1931a). *A Puzzle-Mine.* London: Thomas Nelson and Sons. Puzzles collected from the works of the late H. E. Dudeney by J. Travers.

Dudeney, Henry E. (1931b). *Puzzles and Curious Problems.* London: Thomas Nelson and Sons.

Dudeney, Henry E. (1967). *536 Puzzles & Curious Problems.* New York: Charles Scribner's Sons. Combination of *Puzzles and Curious Problems* and *Modern Puzzles and How to Solve Them,* edited by Martin Gardner.

Dudeney, Henry E. (*Dispatch*). Puzzles and prizes. Column in *Weekly Dispatch,* April 19, 1896–December 26, 1903. Initially the author was identified only as the "Dispatch Sphinx" and later as "Henry E. Dudeney (Sphinx)." (1899a): February 26; (1899b): December 17; (1900a): April 8; (1900b): June 3; (1900c): August 26; (1902a): April 6, p. 13; (1902b): April 20, p. 13; (1902c): May 4, p. 13; (1902d): September 7; (1902e): November 16; (1903a): March 15.

Dudeney, Henry E. (*London*). The Canterbury puzzles. Irregular series in *London Magazine,* January 1902–November 1903. (1902a): February, p. 54; (1902b): June, p. 481; (1903a): November, p. 443.

Dudeney, Henry E. (*Strand*). Perplexities. Monthly puzzle column in *The Strand Magazine* from 1910 to 1930. vols. 39–79. Usually the column would be expanded and titled differently for the December issues. (1911a): vol. 41, p. 746; (1911b): vol. 42, p. 108; (1920a): vol. 60, pp. 368, 452; (1923a): vol. 65, p. 405; (1924a): vol. 68, p. 214; (1926a): vol. 71, p. 208; (1926b): vol. 71, p. 522; (1926c): vol. 72, p. 212; (1926 d): vol. 72, p. 316; (1927a): vol. 73, pp. 420; (1927b): vol. 73, pp. 420, 526; (1927c): vol. 74, p. 200; (1929a): vol. 77, p. 602.

Eaton, Walter P. (1911). My fifty years in puzzleland: Sam Loyd and his ten thousand brain-teasers. *The Delineator* (April), 274, 328.

Education Development Center (1997). *The Cutting Edge: Congruence, Area, and Proof.* Dedham, Massachusetts: Janson Publications. Connected Geometry project.

Elliott, C. S. (1982-1983). Some new geometric dissections. *Journal of Recreational Mathematics* **15**(1), 19–27. Baywood Publishing.

Elliott, C. S. (1985-1986). Some more geometric dissections. *Journal of Recreational Mathematics* **18**(1), 9–16. Baywood Publishing.

Eves, Howard (1963). *A Survey of Geometry,* Volume I. Boston: Allyn & Bacon.

Field, J. V. (1979). Kepler's star polyhedra. *Vistas in Astronomy* **23**, 109–41.

Firby, P. A. and C. F. Gardiner (1982). *Surface Topology.* Chichester: Ellis Horwood.

Fiske, Daniel W. (1859). *The Book of the First American Chess Congress.* New York: Rudd & Carleton. The Congress met in New York City in October and November 1857.

Fourrey, E. (1907). *Curiosités Géométriques.* Paris: Vuibert et Nony.

Frederickson, Greg N. (1974). More geometric dissections. *Journal of Recreational Mathematics* **7**(3), 206-12. Baywood Publishing.

Frederickson, Greg N. (1972a). Appendix H. In Harry Lindgren, *Recreational Problems in Geometric Dissections and How to Solve Them.* New York: Dover Publications.

Frederickson, Greg N. (1972b). Assemblies of twelve-pointed stars. *Journal of Recreational Mathematics* **5**(2), 128-32. Baywood Publishing.

Frederickson, Greg N. (1972c). Polygon assemblies. *Journal of Recreational Mathematics* **5**(4), 255-60. Baywood Publishing.

Frederickson, Greg N. (1972d). Several star dissections. *Journal of Recreational Mathematics* **5**(1), 22-6. Baywood Publishing.

Frederickson, Greg N. (1978-1979). Several prism dissections. *Journal of Recreational Mathematics* **11**(3), 161-75. Baywood Publishing.

Frederickson, Greg N. (1996-1997). Maltese Crosses. *Journal of Recreational Mathematics* **28**(2), 111-15. Baywood Publishing.

Freese, Ernest I. (1957). Blueprints containing over two hundred plates. Described briefly in letters by C. Dudley Langford and Dorman Luke. Complete copy not located.

Gailliard, J. (1985). *Bruges et le Franc, ou leur magistrature et leur noblesse.* Brugge. dl. 3. Source for Paul Busschop.

Gardner, Martin (1956). *Mathematics Magic and Mystery.* New York: Dover. See Chapter 8.

Gardner, Martin (Ed.) (1959). *Mathematical Puzzles of Sam Loyd, Volume 1.* New York: Dover Publications.

Gardner, Martin (Ed.) (1960). *Mathematical Puzzles of Sam Loyd, Volume 2.* New York: Dover Publications.

Gardner, Martin (1961). *The 2nd Scientific American Book of Mathematical Puzzles & Diversions.* New York: Simon & Schuster.

Gardner, Martin (1966). *New Mathematical Diversions from Scientific American.* New York: Simon and Schuster.

Gardner, Martin (1969). *The Unexpected Hanging and Other Mathematical Diversions.* Chicago: University of Chicago Press. Updated with a new afterword and an expanded bibliography, 1991.

Gardner, Martin (1986). *Knotted Doughnuts and Other Mathematical Entertainments.* New York: W. H. Freeman & Co.

Gauss, Carl F. (1900). *Werke,* Volume 8. Göttingen: Königliche Gesellschaft der Wissenschaften zu Göttingen. Contains letter from Christian Ludwig Gerling to C. F. Gauss, April 15, 1844. See pp. 242-3, and the series of letters on pp. 240-9.

Gerwien, P. (1833). Zerschneidung jeder beliebigen Anzahl von gleichen geradlinigen Figuren in dieselben Stücke. *Journal für die reine und angewandte Mathematik (Crelle's Journal)* **10**, 228-34 and Taf. III.

Goldberg, Michael (1940). Solution to problem E400. *American Mathematical Monthly* **47**, 490-1.

Goldberg, Michael (1952). Problem E972: Six piece dissection of a pentagon into a triangle. *American Mathematical Monthly* **59**, 106-7.

Goldberg, Michael (1958). Tetrahedra equivalent to cubes by dissection. *Elemente der Mathematik* **13**, 107-9.

Goldberg, Michael (1966). A duplication of the cube by dissection and a hinged linkage. *Mathematical Gazette* **50**, 304-5.

Goldberg, Michael (1969). Two more tetrahedra equivalent to cubes by dissection. *Elemente der Mathematik* **24**, 130-2.

Goldberg, Michael (1974). New rectifiable tetrahedra. *Elemente der Mathematik* **29**, 85-9.

Graves, Robert P. (1889). *Life of Sir William Rowan Hamilton*, Volume III. Dublin: Hodges, Figgis, & Co. Contains letter from Augustus deMorgan to W. R. Hamilton, September 5, 1855. See pp. 499-503.

Gross, Jonathan L. and Thomas W. Tucker (1987). *Topological Graph Theory*. New York: John Wiley.

Grünbaum, Branko and G. C. Shephard (1987). *Tilings and Patterns*. New York: W. H. Freeman & Co.

Hadwiger, H. and P. Glur (1951). Zerlegungsgleichheit ebener Polygone. *Elemente der Mathematik* **6**, 97-106.

Hanegraaf, Anton (1989). The Delian altar dissection. Elst, the Netherlands. Self-published. First booklet in his projected series *Polyhedral Dissections*.

Hart, Harry (1877). Geometrical dissections and transpositions. *Messenger of Mathematics* **6**, 150-1.

Haüy, René J. (1801). *Traité de minéralogie*. Paris: Bachelier. See Volume 1, pp. 29-30, and Atlas, plate II, Fig. 10.

Hilbert, David (1900). Mathematische Probleme. *Nachrichten von der Gesellschaft der Wissenschaften zu Göttingen, Mathematisch-Physikalische Klasse*. Subsequently in *Bulletin of the American Mathematical Society* **8** (1901-1902), 437-79.

Hill, M. J. M. (1895-1896). Determination of the volumes of certain species of tetrahedra without employment of the method of limits. *Proceedings of the London Mathematical Society* **27**, 39-53.

Hoffmann, Professor (1893). *Puzzles Old and New*. London: F. Warne.

Hogendijk, Jan P. (1984). Greek and Arabic constructions of the regular heptagon. *Archive for History of Exact Sciences* **30**, 197-330.

Hooper, William (1774). *Rational Recreations*. London.

Hunger, Rudolf (1921). Anschauliche Beweise für den erweiterten pythagoreischen Lehrsatz. *Zeitschrift für mathematischen und naturwissenschaftlichen Unterricht* **52**, 160-7.

Hutton, Charles (1840). *Recreations in Mathematics and Natural Philosophy*. London. See "Preface" for a biography of Montucla.

Jackson, John (1821). *Rational Amusement for Winter Evenings*. London. Subtitle: "*A Collection of above 200 Curious and Interesting Puzzles and Paradoxes relating to Arithmetic, Geometry, Geography, &c.*" See p. 84 and plate I, Fig. 6.

Jay, Robert (1987). *The Trade Card in Nineteenth-Century America*. Columbia: University of Missouri Press.

Jessen, Børge (1968). The algebra of polyhedra and the Dehn–Sydler theorem. *Mathematica Scandinavica* **22**, 241-56. See p. 242.

Juel, C. (1903). Om endelig ligestore Polyedre. *Nyt Tidsskrift for Mathematik* **14**, 53-63. See p. 55.

Kaufman-Rosen, Leslie (1995). Big Blue's Butterfly. *Newsweek* (March 20), 46.

Kelland, Philip (1855). On superposition. *Transactions of the Royal Society of Edinburgh* **21**, 271-3 and plate V.

Kelland, Philip (1864). On superposition. Part II. *Transactions of the Royal Society of Edinburgh* **33**, 471-3 and plate XX.

Kepler, Johannes (1940). *Gesammelte Werke*. München: C. H. Beck'sche Verlagsbuchhandlung. Edited by Max Caspar.

Keynes, J. M. (1937). William Herrick Macaulay. *The Cambridge Review* (Jan. 15), 169-70.

Kordemsky, Boris A. (1972). *The Moscow Puzzles*. New York: Charles Scribner's Sons. Translation by Albert Parry.

Kraitchik, Maurice (1942). *Mathematical Recreations*. New York: Norton.

Kürschák, Josef (1899). Über das regelmässige Zwölfeck. *Mathematische und naturwissenschaftliche Berichte aus Ungarn* **15**, 196-7.

Lamb, H. (1928). Henry Martyn Taylor. *Proceedings of the Royal Society of London, Series A* **117**, xxix-xxxi.

Langford, C. Dudley (1956). To pentasect a pentagon. *Mathematical Gazette* **40**, 218.

Langford, C. Dudley (1960). Tiling patterns for regular polygons. *Mathematical Gazette* **44**, 105-10.

Langford, C. Dudley (1967a). On dissecting the dodecagon. *Mathematical Gazette* **51**, 141-2.

Langford, C. Dudley (1967b). Polygon dissections. *Mathematical Gazette* **51**, 139-41.

Le Lionnais, François (1983). *Les nombres remarquables.* Paris: Hermann.

Lemon, Don (1890). *The Illustrated Book of Puzzles.* London: Saxon.

Lenhard, H.-C. (1962). Über fünf neue Tetraeder, die einem Würfel äquivalent sind. *Elemente der Mathematik* **17**, 108-9.

Lindgren, H. (1951). Geometric dissections. *Australian Mathematics Teacher* **7**, 7-10.

Lindgren, H. (1953). Geometric dissections. *Australian Mathematics Teacher* **9**, 17-21, 64.

Lindgren, H. (1956). Problem E1210: A dissection of a pair of equilateral triangles: Solution. *American Mathematical Monthly* **63**, 667-8.

Lindgren, H. (1957). Problem E1240: Two six-piece dissections. *American Mathematical Monthly* **64**, 368-9.

Lindgren, H. (1958). Problem E1309: Dissection of a regular pentagram into a square. *American Mathematical Monthly* **65**, 710-11.

Lindgren, H. (1960). A quadrilateral dissection. *Australian Mathematics Teacher* **16**, 64-5.

Lindgren, H. (1961). Going one better in geometric dissections. *Mathematical Gazette* **45**, 94-7.

Lindgren, H. (1962a). Dissecting the decagon. *Mathematical Gazette* **46**, 305-6.

Lindgren, Harry (1962b). Puzzles and problems 1. A solid dissection problem. *Recreational Mathematics Magazine* (12), December, 23.

Lindgren, Harry (1962c). Three Latin-cross dissections. *Recreational Mathematics Magazine* (8), April, 18-19.

Lindgren, H. (1964a). *Australian Mathematics Teacher* **20**, 52-4.

Lindgren, Harry (1964b). *Geometric Dissections.* Princeton, N.J.: D. Van Nostrand Company.

Lindgren, Harry (1968). Some approximate dissections. *Journal of Recreational Mathematics* **1**(2), 79-92. Baywood Publishing.

Lindgren, Harry (1970). A dissection problem by Sam Loyd. *Journal of Recreational Mathematics* **3**(1), 54-5. Baywood Publishing.

Lindgren, Harry (1972). *Recreational Problems in Geometric Dissections and How to Solve Them.* New York: Dover Publications. Greg Frederickson revised (Lindgren 1964b) by adding a new appendix (Appendix H) and updating two other appendices.

Lindgren, Judy (1992). Meny years trying to reform spelling. *Canberra Times*, Midweek Magazine (July 8), p. 23. Obituary for Harry Lindgren.

Lines, L. (1935). *Solid Geometry.* London: Macmillan & Co. Reprinted by Dover Publications, 1965.

Lipton, L. (1959). *The Holy Barbarians.* New York: Messner.

Lowry, Mr. (1814). Solution to question 269, [proposed] by Mr. W. Wallace. In T. Leybourn (Ed.), *Mathematical Repository*, Volume III, pp. 44-6 of Part I. London: W. Glendinning. Leybourn, Lowry, and Wallace were all with the Royal Military College, Sandhurst.

Loyd, Sam (1914). *Cyclopedia of Puzzles.* New York: Franklin Bigelow Corporation. Consisting of the contents of the four issues of *Our Puzzle Magazine*, which was published quarterly by Sam Loyd, Sr., starting in June 1907.

Loyd, Sam (1928). *Sam Loyd and His Puzzles.* New York and Newark, N.J.: Barse & Co.

Loyd, Sam (*Eagle*). Loyd's Puzzles. Puzzle column in Sunday edition of *Brooklyn Daily Eagle* from March 22, 1896 to April 25, 1897. (1896a): April 12, p. 20; (1897a): Feb. 14 and 21, p. 26.

Loyd, Sam (*Eliz. J.*). Puzzle column in *Elizabeth Journal.* (1908a): Oct. 29.

Loyd, Sam (*Home*). Sam Loyd's Own Puzzle Page. Monthly puzzle column in *Woman's Home Companion*, 1903–1911. Originally entitled "Mr. X's ???," then "Do You Know Mr. X ???," then "The Xcentric Mr. X: His Page," then "Professor X – Sam Loyd Puzzle Trust." (1908a): November, p. 51.

Loyd, Sam (*Inquirer*). Mental Gymnastics. Puzzle column in Sunday edition of *Philadelphia Inquirer*, October 23, 1898–1901. Originally entitled "Mental Gymnastics for Young and Old," the column usually appeared in the Colored Section and included problems in chess and whist. (1898a): November 6; (1899a): February 26; (1900a): December 9; (1901a): February 24; (1901b): April 14; (1901c): May 26; (1901d): July 21; (1901e): August 11.

Loyd, Sam (*Press*). Sam Loyd's Puzzles. Puzzle column in Sunday edition of *Philadelphia Press*, February 23–June 29, 1902. The column typically appeared on the third page of a four-page section containing comics and cartoons. (1902a): May 5; (1902b): June 29.

Loyd, Sam (*Tit-Bits*). Weekly puzzle column in *Tit-Bits*, starting October 3, 1896, and continuing into 1897. Dudeney, under the pseudonym "Sphinx," wrote commentary and handled the awarding of prize money. Later in 1897 Dudeney took over the column. (1897a): April 24, p. 59 (1897b): May 15, p. 117 (1897c): June 26, p. 233.

Lucas, Édouard (1883). *Récréations Mathématiques*, Volume 2. Paris: Gauthier-Villars. Second of four volumes. Second edition (1893) reprinted by Blanchard in 1960. See pp. 151 and 152 in Volume 2 of this edition.

Macaulay, W. H. (1914). The dissection of rectilineal figures. *Mathematical Gazette* **7**, 381–8.

Macaulay, W. H. (1915). The dissection of rectilineal figures. *Mathematical Gazette* **8,** 72–6 and 109–15.

Macaulay, W. H. (1919a). The dissection of rectilineal figures. *Messenger of Mathematics* **48**, 159–65.

Macaulay, W. H. (1919b). The dissection of rectilineal figures (continued). *Messenger of Mathematics* **49**, 111–21.

Macaulay, W. H. (1922). The dissection of rectilineal figures (continued). *Messenger of Mathematics* **52**, 53–6.

MacMahon, Percy A. (1921). *New Mathematical Pastimes.* London: Cambridge University Press. Reprinted 1930. See Fig. 135a on p. 113.

MacMahon, Percy A. (1922). Pythagoras's theorem as a repeating pattern. *Nature* **109**, 479.

Mahlo, Paul (1908). *Topologische Untersuchungen über Zerlegung in ebene und sphaerische Polygone.* Halle, Germany: C. A. Kaemmerer. Doctoral dissertation for the Vereinigte Friedrichs-Universität in Halle-Wittenberg. See pp. 13, 14 and Fig. 7.

Maxwell, E. A. (1969). C. Dudley Langford. *Mathematical Gazette* **53**, 314.

Milne, E. A. and F. P. White (1949). Herbert William Richmond. *Mathematical Gazette* **24**, 68–80.

Moser, Leo (1949). Problem E860: A volume dissection. *American Mathematical Monthly* **56**, 694.

Mott-Smith, Geoffrey (1946). *Mathematical Puzzles for Beginners and Enthusiasts.* Philadelphia: Blakiston Co. Reprinted by Dover Publications, New York, 1954.

Needham, Joseph (1959). *Science and Civilisation in China*, Volume *3: Mathematics and the Sciences of the Heavens and the Earth.* Cambridge: Cambridge University Press. See Fig. 52 on p. 29, and discussion on p. 27.

Newing, Angela (1994). Henry Ernest Dudeney: Britain's greatest puzzlist. In R. K. Guy and R. E. Woodrow (Eds.), *The Lighter Side of Mathematics*, pp. 294–301. Washington, D.C.: The Mathematical Association of America.

O'Beirne, Thomas H. (1961). A six-block cycle for six step-cut pieces. *New Scientist* (224), March 2, 560-1. In the weekly column "Puzzles and Paradoxes."

Ore, Oystein (1953). *Cardano, The Gambling Scholar*. Princeton, N.J.: Princeton University Press.

Ozanam, Jacques (1778). *Récréations Mathématiques et Physiques*. Paris: Claude Antoine Jombert, fils. See figures 123-126 and pages 297-302. This is material added by Jean Montucla, who is listed as a reviser under the pseudonym of M. de Chanla. See also pages 127-9 of *Recreations in Mathematics and Natural Philosophy*, by Jacques Ozanam, London, Thomas Tegg, 1840, translated from Montucla's edition by Charles Hutton, with additions.

Panckoucke, André-J. (1749). *Les Amusemens Mathématiques*. Lille: Chez André-Joseph Panckoucke.

Paterson, David (1988). Two dissections in 3-D. *Journal of Recreational Mathematics* **20**(4), 257-70. Baywood Publishing.

Paterson, David (1989). T-dissections of hexagons and triangles. *Journal of Recreational Mathematics* **21**(4), 278-91. Baywood Publishing.

Paterson, David (1995a). Dissections of squares, unpublished manuscript.

Paterson, David (1995b). Enneagon dissections, unpublished manuscript. To appear in *Journal of Recreational Mathematics*.

Paterson, David (1995c). Geometric dissections in 4-D, unpublished manuscript. To appear in *Journal of Recreational Mathematics*.

Perigal, F. (1901). *Henry Perigal F.R.A.S., &c.: A Short Record of his Life and Works*. London: Bowles & Sons.

Perigal, Henry (1873). On geometric dissections and transformations. *Messenger of Mathematics* **2**, 103-5.

Perigal, Henry (1875). Geometrical dissections and transformations, no. II. *Messenger of Mathematics* **4**, 103-4.

Perigal, Henry (1891). *Graphic Demonstrations of Geometric Problems*. London: Bowles & Sons. On cover: "Association for the Improvement of Geometrical Teaching. Geometric Dissections and Transpositions." (The association was later renamed Mathematical Association.)

Reid, Robert (1987). Disecciones geometricas. *Umbral* (2), 59-65. Published in Lima, Peru, by Asociacion Civil Antares. The author's name as listed in the article is Robert Reid Dalmau, conforming to Spanish custom, but is listed here in the form that Robert prefers.

Richmond, H. W. (1943). Note 1672: A geometrical problem. *Mathematical Gazette* **27**(275), 142.

Richmond, H. W. (1944). Note 1704: Solution of a geometrical problem. *Mathematical Gazette* **28**(278), 31-2.

Roberts, David L. (1996). Albert Harry Wheeler (1873-1950): A case study in the stratification of American mathematical activity. *Historia Mathematica* **23**, 269-87.

Rosenbaum, Joseph (1947). Problem E721: A dodecagon dissection puzzle. *American Mathematical Monthly* **54**, 44.

Sager, Ira (1995). The butterfly: From a little girl's building blocks. *Business Week* (July 24), 72.

Santaniello, A. E. (1970). Introduction. In *The Book of Architecture*. New York: Benjamin Blom. Introduction to facsimile of translation of Serlio by Peake.

Sayili, Aydin (1958). Sābit ibn kurranin Pitagor teoremini temini. *Türk Tarih Kurumu. Bulleten* **22**, 527-49.

Sayili, Aydin (1960). Thâbit ibn Qurra's generalization of the Pythagorean theorem. *Isis* **51**, 35-7.

Schlömilch, O. (1868). Ein geometrisches Paradoxon. *Zeitschrift für Mathematik und Physik* **13**, 162.

Schmerl, James (1973). Problem #240: A Pythagorean dissection. *Journal of Recreational Mathematics* **6**(4), 315-16. Baywood Publishing. See also **7**(2), 153, 1974.

Schöbi, Philipp (1985). Ein elementarer und konstruktiver Beweis für die Zerlegungsgleichheit der Hill'schen Tetraeder mit einem Quader. *Elemente der Mathematik* **40**, 85-97.

Serlio, Sebastiano (1551). *Il primo-quinto libro d'architettura*. Venetia. Translated from Italian to Dutch by Pieter Coecke, then from Dutch to English by Robert Peake for the edition *The Book of Architecture*, published by Peake in London in 1611. Facsimile printed by Benjamin Blom, New York, in 1970. See the first book, *Of Geometrie*, Fol. 13.

Sherie, Fenn (1926). The Puzzle King: An interview with Henry E. Dudeney. *The Strand Magazine* (71), 398-404.

Singmaster, David (1995). *Sources in Recreational Mathematics: An Annotated Bibliography*. London: South Bank University. 7th preliminary edition.

Slocum, Jerry (1996). Puzzle cards. Color reproduction of eight advertising puzzle cards from the late nineteenth century, with commentary. The cards are from the Jerry Slocum Puzzle Collection.

Slocum, Jerry and Jack Botermans (1987). *Puzzles Old and New*. Seattle: University of Washington Press. See p. 144.

Slocum, Jerry and Jack Botermans (1994). *The Book of Ingenious & Diabolical Puzzles*. New York: Time Books. See p. 11.

Steinhaus, H. (1960). *Mathematical Snapshots* (2nd ed.). New York: Oxford University Press. See p. 11.

Sturm, Johann C. (1700). *Mathesis Enumerata: or, the Elements of the Mathematicks*. London: Robert Knaplock. Translation (by J. Rogers?) of 1695 work *Mathesis Enumerata*. See pp. 20-1 and Fig. 29.

Sydler, J.-P. (1956). Sur les tétraèdres équivalents à un cube. *Elemente der Mathematik* **11**, 78-81.

Sydler, J.-P. (1965). Conditions nécessaires et suffisantes pour l'équivalence des polyèdres de l'espace euclidien à trois dimensions. *Commentarii Mathematici Helvetica* **40**, 43-80.

Taylor, H. M. (1905). On some geometrical dissections. *Messenger of Mathematics* **35**, 81-101.

Tilson, Philip G. (1978-1979). New dissections of pentagon and pentagram. *Journal of Recreational Mathematics* **11**(2), 108-11. Baywood Publishing.

Tjebbes, T. (1969). ABT Abstract, nu concreet. *ABT-mededelingen* **1**(2), 1-6. Photoreport of an exposition in Arnheim. See photograph on lower left of page 6.

Travers, J. (1933). Puzzles and problems. *The Education Outlook*, 145.

Travers, J. (1952). *Puzzling Posers*. London: George Allen & Unwin Ltd.

van Delft, Pieter and Jack Botermans (1978). *Creative Puzzles of the World*. New York: Harry N. Abrams.

Varsady, Alfred (1985-1986). The dissection of sets of polygons. *Journal of Recreational Mathematics* **18**(4), 256-68. Baywood Publishing.

Varsady, Alfred (1989). Some new dissections. *Journal of Recreational Mathematics* **21**(3), 203-9. Baywood Publishing.

Wallace, William (Ed.) (1831). *Elements of Geometry* (8th ed.). Edinburgh: Bell & Bradfute. First six books of Euclid, with a supplement by John Playfair.

Wells, David (1975). Figures: On gems and generalisations. *Games & Puzzles* (38), June, 40. Author not identified, but attributed to David Wells in (Wells 1991).

Wells, David (1991). *The Penguin Dictionary of Curious and Interesting Geometry*. London: Penguin Books.

Wenninger, Magnus J. (1971). *Polyhedron Models*. London: Cambridge University Press.

Wheeler, A. H. (1935). Problem E4. *American Mathematical Monthly* **42**, 509-10.

White, Alain C. (1913). *Sam Loyd and His Chess Problems*. Leeds, England: Whitehead and Miller.

Woepcke, F. (1855). Analyse et extrait d'un recueil de constructions géométriques par Aboûl Wafâ. *Journal Asiatique* **V**, 318-59.

301

Index of Dissections

Dissections are ordered by the following conventions:

1. Two-dimensional dissections precede three-dimensional dissections.

2. The two-dimensional figures are ordered with those represented by letters (in alphabetical order) following those represented by $\{p\}$ or $\{p/q\}$ (in lexicographic order on (p,q)).

3. A dissection is listed under the figure it involves that comes latest in the list.

4. For a figure such as $\{r\}$, dissections of it involving $\{p\}$ or $\{p/q\}$ with $p < r$ come first. Within dissections involving only $\{r\}$, special relationships come first (ordered lexicographically), then general relationships (ordered lexicographically), and then a $\{r\}$s to b $\{r\}$s (ordered lexicographically on (a,b)).

$\{3\}$
 for $1^2 + (\sqrt{3})^2 = 2^2$, 42
 $1^2 + 2^2 + 2^2 = 3^2$, 45
 for $1^2 + 3^2 + 3^2 + 9^2 = 10^2$, 96
 $2^2 + 3^2 + (\sqrt{12})^2 = 5^2$, 45
 for $3^2 + 4^2 + 12^2 = 13^2$, 95
 for $3^2 + 4^2 = 5^2$, 91
 for $5^2 + 12^2 = 13^2$, 92
 for $7^2 + 6^2 + 6^2 = 11^2$, 96
 for $9^2 + 8^2 + 12^2 = 17^2$, 95
 for $15^2 + 8^2 = 17^2$, 92
 for $x^2 + y^2 + z^2 + w^2 = v^2$
  $w = z = y$, 47
 for $x^2 + y^2 + z^2 = w^2$
  $x + y = w$, 46
 for $x^2 + y^2 + z^2 = w^2$
  $z = y$, 44
 two to one, 41
 three equal to one, 43, 55
 three unequal to one, 44
 five to one, 56
 seven to one, 55
 thirteen to three, 57
 $a^2 + ab + b^2$ to one, 55

$\{\tilde{3}\}$
 to $\{\tilde{3}\}$, 139
$\{4\}$
 to $\{3\}$, 136
 $\{\tilde{3}\}$, 224
 for $1^2 + (\sqrt{3})^2 + (\sqrt{5})^2 = 3^2$, 37
 for $1^2 + 2^2 + 2^2 = 3^2$, 80
 for $1^2 + 4^2 + 8^2 = 9^2$, 80
 for $1^2 + n^2 + n^2 + (n^2)^2 = (n^2 + 1)^2$, 88
 for $2^2 + (\sqrt{8})^2 + (\sqrt{13})^2 = 5^2$, 37
 $2^2 + 3^2 + (\sqrt{3})^2 = 4^2$, 37
 for $2^2 + 3^2 + 6^2 = 7^2$, 71
 for $2^2 + 4^2 + 5^2 + 6^2 = 9^2$, 87
 for $3^2 + 4^2 = 5^2$, 62
 for $3^2 + 12^2 + 4^2 = 13^2$, 84
 for $4^2 + 9^2 + 48^2 = 49^2$, 87
 for $5^2 + 12^2 = 13^2$, 64, 65
 for $5^2 + 14^2 + 2^2 = 15^2$, 82
 for $6^2 + 7^2 = 9^2 + 2^2$, 76
 for $6^2 + 17^2 = 18^2 + 1^2$, 75
 for $7^2 + 4^2 = 8^2 + 1^2$, 77
 for $7^2 + 6^2 + 6^2 = 11^2$, 86
 for $7^2 + 24^2 = 25^2$, 65
 for $8^2 + 1^2 + 4^2 = 9^2$, 83
 for $9^2 + 8^2 + 12^2 = 17^2$, 84
 for $9^2 + 12^2 + 20^2 = 25^2$, 87

{4} (*cont.*)

for $9^2 + 32^2 + 24^2 = 41^2$, 84

for $10^2 + 5^2 = 11^2 + 2^2$, 74

for $10^2 + 23^2 = 25^2 + 2^2$, 75

for $12^2 + 15^2 + 16^2 = 25^2$, 80

for $13^2 + 18^2 = 22^2 + 3^2$, 79

for $15^2 + 8^2 = 17^2$, 68

for $16^2 + 18^2 + 24^2 = 34^2$, 84

for $17^2 + 6^2 = 15^2 + 10^2 = 18^2 + 1^2$, 78

for $23^2 + 24^2 = 32^2 + 9^2$, 78

for $34^2 + 12^2 + 12^2 = 38^2$, 86

for $35^2 + 12^2 = 37^2$, 68

for $36^2 + 2^2 + 12^2 = 38^2$, 83

for $x^2 + y^2 + z^2 = w^2$

$w = x + y$, 36

$w + x = y + z$, 37

$x^2 + y^2 + z^2$, 35

$z = 2(w - y)$, 36

$z = y$, 35

$z = y$ and $x = \sqrt{2}y$, 36

$z^2 = y^2 + (w - x)^2$, 37

special relationships, 35

two equal to one, 30

two unequal squares to one, 28, 30

two unequal to 2 unequal, 30

two unequal attached to 1, 30

three equal to one, 31

three unequal to one, 33

five to two, 53

13 to 1, 52

$a^2 + b^2$ to one, 52

$a^2 + b^2$ to $c^2 + d^2$, 53

{$\tilde{4}$}

to {$\tilde{4}$}, 120

to {$\tilde{3}$}, 139

{5}

to {3}, 140

to {4}, 120

for $(\sqrt{4 - \phi})^2 + \phi^2 = (\phi\sqrt{5})^2$, 217

for $1^2 + \phi^2 + (\phi^2)^2 = (2\phi)^2$, 215

for $5^2 + 12^2 = 13^2$, 97

four to one, 23

five to one, 197

six to one, 58

{$\mathring{5}$}

two similar to one, 226

two similar to two, 227

{⑤}

two similar to one, 225

{5/2}

to {4}, 127

to {5}, 146

$(\sin\frac{2\pi}{5})^2 + (\cos\frac{2\pi}{5})^2 = 1^2$, 217

for $1^2 + \phi^2 + (\phi^2)^2 = (2\phi)^2$, 215

for $(2\phi^3)$-{5/2} and {5} to $(\phi^3\sqrt{5})$-{5}, 215

five to one, 198

{6}

to {3}, 142

to two {3}s, 180

two to {3}, 180

to {4}, 118

to {5}, 284

for $1^2 + (\sqrt{3})^2 = 2^2$, 47

for $1^2 + 6^2 = (\sqrt{37})^2$, 104

for $1^2 + (\sqrt{48})^2 = 7^2$, 102

for $1^2 + (\sqrt{63})^2 = 8^2$, 103

for $2^2 + 3^2 = (\sqrt{13})^2$, 48

for $2^2 + 3^2 + 6^2 = 7^2$, 99

for $(\sqrt{7})^2 + 3^2 = 4^2$, 48

for $3^2 + 4^2 = 5^2$, 98

for $5^2 + 12^2 = 13^2$, 98

two unequal to one, 47

seven to one, 57

$a^2 + ab + b^2$ to one, 57

{$\tilde{6}$}

two to one, 228

{$\tilde{6}$} (notched rectangle)

four to one, 229

{6/2}

to {3}, 110

three to {3}, 208

to {4}, 122

to {6}, 153

to two {6}s, 178

two to {6}, 179

for $1^2 + (\sqrt{3})^2 = 2^2$, 50

for $(\sqrt{3})^2 + 2^2 = (\sqrt{7})^2$, 50

for $1^2 + 2^2 + 2^2 = 3^2$, 100

for $3^2 + 4^2 = 5^2$, 99

two unequal to one, 48, 49

three to one, 192

seven to one, 203

{7}

to {4}, 128

for $1^2 + \vartheta^2 + (\vartheta^2 - 1)^2 = (\vartheta^2 + \vartheta - 2)^2$, 220

two to one, 206

seven to one, 205

eight to one, 206

{7/2}

$(\sin\frac{2\pi}{7})^2 + (\cos\frac{2\pi}{7})^2 = 1^2$, 218

{7/3}

to {7}, 212

to two {7}s, 1, 183, 186

two to {7}, 183

{8}

to {3}, 147

to {4}, 150

to {6}, 133

{8} (*cont.*)
 two to one, 188
 three to one, 196
 three to two, 201
 five to one, 199
 six to one, 201
{8/2}
 to {4}, 109, 175
 two to one, 188
 three to one, 196
 five to one, 200
{8/3}
 to {3}, 148
 to {4}, 154
 to {6/2}, 148
 to {8}, 177
 two to one, 189
 two unequal to one, 50
{9}
 to {3}, 145
 to {4}, 131
 to {6}, 131
 three to one, 192
 seven to one, 204
{9/3}
 to {6}, 183
{9/4}
 to {9/2}, 180
{10}
 to {3}, 149
 to {4}, 134
 to {5/2}, 173, 185
 to 2 {5}s and 1 {5/2}s, 213
 to 2 {5}s and 2 {5/2}s, 212
 to 6 {5}s and 2 {5/2}s, 214
 to 8 {5}s and 3 {5/2}s, 212
 to 16 {5}s and 5 {5/2}s, 214
 to 22 {5}s and 7 {5/2}s, 213
 five to one, 198
{10/2}
 to {5}, 112
{10/3}
 to two {5/2}s, 174
 to two {10}s, 179
 two to {10}, 179
 to ϕ^3-{5/2} and {5/2}, 214
{10/4}
 to two {5/2}s, 182
 to {10/3}, 179
{12}
 to {4}, 107
 to three {4}s, 3
 to {6}, 152
 to {6/2}, 284
 two to one, 189

 three to one, 193
 three to two, 202
 five to one, 200
 six to one, 201
{12/2}
 to {3}, 182
 to {4}, 134
 to {6}, 110
 two to {6}, 209
 two to three {6}s, 209
 to {6/2}, 178
 to two {6/2}s, 208
 to three {6/2}s, 210
 three to two {6/2}s, 210
 two to one, 189
 three to one, 194
 three to two, 203
{12/3}
 to {4}, 109
 two to one, 191
 three to one, 194
{12/4}
 to {4}, 155
 two to one, 191
 three to one, 195
{12/5}
 to {4}, 182
 to {12}, 178
 to two {12}s, 177
 two to {12}, 177
{16/2}
 to {8}, 176
{C}
 to {4}, 167
{G}
 to {3}, 143
 to {4}, 105
 to {C}, 167
 to {C'}, 168
 for $1^2 + 2^2$ to square, 24
 for $1^2 + 2^2 + 2^2 = 3^2$, 101
 for $3^2 + 4^2 = 5^2$, 101
 for $15^2 + 8^2 = 17^2$, 101
 two to one, 54
 three to one, 123
 five to one, 54
 five to two, 54
{L}
 to {8}, 284
 to {12}, 107
 to {12/5}, 211
 to {G}, 144
{L'}
 to {4}, 114
 to {G}, 114

305

{M}
   to {G}, 161, 162
   to {4}, 157
{M′}
   to {4}, 159
{M‴}
   to {4}, 162
{M̆}
   to {4}, 156
{O}
   two to {D}, 163
{O′}
   two to {D}, 165
(10,12)-platform
   to (13,14)-platform, 257
{8/2}-prism
   two to one, 259
{10/4}-prism
   one to two {5/2}-prisms, 261
{12}-prism
   3-high to 2-high, 266
{12/2}-prism
   three to one, 260
{12/3}-prism
   two to one, 260

{4,3}-surface
   to {3,3}-surface, 246
{3,3}
   irregular to mirror image, 232
{3,3}$_{H1}$
   certain Hill to isosceles right triangle
      prism, 234
   to {3}-prism, 235, 236
{4,3}
   to {3,4} plus 2 {3,3}s, 243
   for $1^3 + 1^3 + 5^3 + 6^3 = 7^3$, 253
   for $1^3 + 6^3 + 8^3 = 9^3$, 251
   for $3^3 + 4^3 + 5^3 = 6^3$, 248
   for $12^3 + 1^3 = 10^3 + 9^3$, 251
   two to one, 239
   two unequal to one, 240
   three to one, 239
{R: $(2 \times 1 \times 1)$}
   to cube, 237
{RD}
   to cube, 242
t{3,4}
   to cube, 241
   two to cube, 241

General Index

advantageous dissection, 212
anchor point of a T-strip, 137
approximate dissections, 276, 277
Archimedean solids, 17
auspicious dissections, 172

Bennett, Geoffrey T.
  biography, 152
  octagon to square, 150
Bradley, Harry C.
  biography, 41
  dissections, 41, 91, 122
Brahmagupta
  solutions to $x^2 + y^2 = z^2$, 62
Brodie, B.
  two similar rectangles, 30
Brodie, Robert
  biography, 121
  pentagon to square, 120
bumpy plain-strip technique, 134
Busschop, Paul
  biography, 119
  hexagon to square, 118
  was improved on, 32, 37, 119

Cadwell, James H.
  biography, 251
  dissections, 251, 252
  was improved on, 251
Cardano, Girolamo
  biography, 61
  step technique, 60
Carpitella, Duilio
  biography, 245
  octahedron plus 2 tetrahedra to cube, 243
Carver, W. B.
  was improved on, 277
Cataneo, Pietro
  explained Serlio's paradox, 272

Charlton, H. G.
  was improved on, 128
Cheney, Wm. Fitch
  was improved on, 236
Cohn, M. J.
  lower bound, 223
  was improved on, 224
Collison, David M.
  biography, 93
  dissections, 55, 91, 92, 95–100, 228
  method: Pythagoras's class, 66
Cossali's class, 83
  method for triangles, 95
Cross and Crescent Puzzle, 165
Crusader's Puzzle, 169
customized strips, 133, 147

Dintzl, Erwin
  biography, 39
  dissection: law of cosines, 39
Diophantus
  derived $7^2 + 4^2 = 8^2 + 1^2$, 73
  discovered $2^2 + 3^2 + 6^2 = 7^2$, 80
  solutions to $x^2 + y^2 = z^2$, 62
dn–left, 91
dn–right, 91
double P-slide, 33
double-square-difference class, 79
Dudeney, Henry E.
  biography, 81
  design of crescent $\{C'\}$, 168
  discussed step technique, 60
  dissections, 45, 54, 64, 65, 80, 270
  square to triangle?, 136
  was improved on, 65, 71, 122, 142, 150, 157
Duffy, Emmet J.
  cubes for $3^3 + 4^3 + 5^3 = 6^3$, 248

Elliott, C. Stuart
  biography, 209

Elliott, C. Stuart (*cont.*)
  dissections, 3, 203, 208
  was improved on, 176, 208-10
enneagon, 10

favorable dissections, 188
Fibonacci numbers, 273
Fibonacci's formula, 73
flip-down step technique, 92
flip-up step technique, 92
Frederickson, Greg N.
  biography, 267
  was improved on, 192, 195, 196, 201
Freese, Ernest Irving
  biography, 154
  dissections, 87, 152, 188
  was improved on, 145, 189, 198

Gauss, Carl Friedrich
  correspondence with Gerling, 231
Gerling, Christian Ludwig
  biography, 231
  tetrahedron to mirror image, 232
Gerwien, P.
  biography, 158
  triangles of equal height, 158
Gilson, Bruce
  was improved on, 154
Goldberg, Michael
  biography, 141
  pentagon to triangle, 141
  was improved on, 141, 237
golden ratio, 212
  and Fibonacci numbers, 273
  pentagon dimensions, 197
Guido Mosaics Puzzle, 62

H-slide, 90, 129
Hanegraaf, Anton
  biography, 243
  dissections, 131, 162, 192, 193, 235, 237,
    239-42
  enneagon strip, 131
  trapezoid slide, 130
  was improved on, 128, 132
Hart, Harry
  biography, 226
  dissections, 42, 225, 226
Hill, A. E.
  Maltese Cross to square, 157
hinged
  cyclicly, 26, 27, 102, 103, 121, 130, 140,
    143, 238
  fully, 25, 31, 106, 110, 136, 139, 140, 143,
    159, 235

partially, 27, 30, 47, 54, 102, 103, 141-3,
    151, 152, 173, 177, 217, 218
  variously, 26, 31, 47, 106, 110, 138, 139,
    141-2, 152, 177, 217, 218
hole dissection, 169
honeycomb, 18
Hooper, William
  geometric money paradox, 271
Hunger, Rudolf
  dissection: law of cosines, 38

IBM butterfly
  expanding keyboard, 68
interchanged step technique, 77
iso-hepta triangles, 218
iso-penta triangles, 212

Jackson, John
  biography, 164
  was improved on, 163
Jessen, Børge
  tetrahedron to mirror image, 232
Juel, C.
  tetrahedron to mirror image, 232

Karidis, John P.
  expanding keyboard, 70
Kelland, Philip
  biography, 33
  dissections, 32, 270
  P-slide, 33

Langford, C. Dudley
  biography, 197
  dissections, 188, 197
  was improved on, 192, 197
law of cosines dissection, 38
Lebesgue's formula, 80
Lemaire, Bernard
  biography, 162
  design of Maltese {M̌}, 155
  dissections, 161, 162
Lemon, Don
  biography, 106
  Greek Cross to square, 105
Lindgren, Harry
  biography, 183
  dissections, 23, 41, 47, 48, 54, 55, 57, 107,
    123, 139, 142-4, 155, 161, 168, 173,
    174, 177-80, 182, 188, 189, 192, 211,
    276, 277, 284
  element for Maltese Cross, 160
  Q-slide, 42
  was improved on, 67, 109, 110, 112, 127,
    128, 131-4, 147, 149, 154, 178, 182,
    190, 193, 201, 229

Loyd, Sam
  biography, 73
  chessboard paradox, 273
  design of crescent {C}, 167
  dissections, 23, 24, 64, 71, 80, 105, 163,
      165, 167
  erroneous dissections, 167, 270
  letterhead, 72
  puzzle cards, 65, 105
  puzzles, 5, 62, 64, 89, 165, 169, 270
  stutter-step technique, 65
  was improved on, 122

Macaulay, William H.
  biography, 123
  superposing tessellations, 29
  quadrilateral to quad., 120
MacMahon, Percy
  superposing tessellations, 29
Mahlo, Paul
  superposing tessellations, 29
McElroy, C. W.
  square to triangle?, 136
Methods 1A–C, 66, 74, 82
Methods 2A–C, 67, 75, 83
Methods 3–6, 76, 78, 79, 85
Methods 7–13, 91, 93, 94, 96, 102–4
Montucla, Jean Etienne
  biography, 223
  dissections, 119, 222
Mott-Smith, Geoffrey
  biography, 111
  dissections, 37, 80, 110
Mrs. Pythagoras' Puzzle, 64

nonorientable surface tessellation, 184–6

O'Beirne, Thomas H.
  biography, 255
  dissections, 248, 255
optimized strip, 133
optimizing strip technique, 124

P-slide, 32
P-strip technique, 119
parhexagon, 90
partial replication symmetry, 23
Paterson, David
  biography, 53
  formula for triangles, 55
  two truncated octahedra to cube, 241
  was improved on, 132, 241
Penta class, 76
Penta-penta class, 78
Perigal, Henry

biography, 31
  dissections, 30, 32
platform
  stepped, 256
Plato
  dissections, 30, 43
Plato's class, 67
  method for triangles, 92
  method for squares, 67
Plato-extended class, 94
polyhedral tessellations, 112, 114, 115
post for 3D strip, 243
PP-double class, 85
  method for triangles, 96
PP-minus class, 75
PP-plus class, 82
prisms, 19
propitious dissection, 208
Pythagoras's class, 62
  method for triangles, 91
  method for squares, 66
Pythagoras-extended class, 93
Pythagoras-minus class, 74
Pythagoras-plus class, 81

Q-slide, 41

rational dissection, 24
Red Spade Puzzle, 169
reflection symmetry, 20
Reid, Robert
  biography, 203
  bumpy plain-strip, 134
  dissections, 50, 52, 79, 84, 86, 87, 100,
      101, 145, 192, 193, 195, 196, 202,
      208–10, 212, 214, 215, 229, 251, 253
  was improved on, 135, 209
  relationships for $x^2 + y^2 + z^2 = w^2$, 35, 44
  relationships for stars, 175, 181
replication symmetry, 22
rhombic dodecahedron, 18
rhombohedron, 233
Richards, Sgt. E. T.
  squares for $7^2 + 24^2 = 25^2$, 65
Richmond, Herbert W.
  biography, 248
  was improved on, 248
Rosenbaum, Joseph
  biography, 190
  was improved on, 189
rotational symmetry, 21

Sankey, C. E. P.
  biography, 129
  was improved on, 128

Schöbi, Philipp
  Hill tetrahedron to {3}-prism, 235
Schlömilch, O.
  chessboard paradox, 272
Schmerl, James
  biography, 99
  hexagons: $3^2 + 4^2 = 5^2$, 98
Sedan Chair Puzzle, 5
semiregular polyhedron, 18
Serlio, Sebastiano
  biography, 273
  table paradox, 271
Sheldon, Eli Lemon
  biography, 106
silvern ratio
  heptagon dimensions, 218
  recurrence relation, 291
slides
  hexagon slide (H-slide), 90
  parallelogram slide (P-slide), 32
  quadrilateral slide (Q-slide), 41
  trapezoid slide (T-slide), 130
Smart Alec Puzzle, 270
square-difference class, 78
square-sum class, 84
square-sum-plus class, 95
star relationships, 174, 181
Steinhaus, H.
  was improved on, 55
step technique, 60
stutter-step technique, 65
superposing tessellations, 29
surface dissection, 245
Sydler, Jean-Pierre
  biography, 234
  Hill tetrahedron to isosceles right triangle
    prism, 234
Szeps
  {L'} to square, 114

T-slide, 130, 238
T-strip technique, 137
Tai Chen
  dodecagon to 3 squares, 3
Taylor, Henry M.
  biography, 140
  triangle to triangle, 139
tessellation, 14
  repetition pattern, 15
  tessellation element, 14
Thābit ibn Qurra
  biography, 29
  two unequal squares to 1, 28
Theobald, Gavin
  biography, 147

customized strips, 133
dissections, 128, 132–5, 145–9, 246
optimized strip, 133
surface dissections, 245
Thewlis, J. H.
  cubes for $3^3 + 4^3 + 5^3 = 6^3$, 248
Tilson, Philip Graham
  pentagram to square, 127
  was improved on, 146
top-versus-bottom symmetry, 259
translation with no rotation, 22
translational, 22
Travers, James
  claim: octagon to square, 152
Tri-minus class, 103
Tri-plus class, 104
Tri-root class, 102
truncated octahedron, 18
turned over pieces, 23

up–left, 91
up–right, 91

Varsady, Alfred
  biography, 220
  dissections, 55, 58, 203, 212–15, 217, 219,
    220
  expansion of iso-penta triangles, 212
  $\sqrt{7}$ in heptagon structure, 205
  $\sqrt{8}$ in heptagon structure, 206
  was improved on, 214, 215

Wafā, Abū'l
  biography, 52
  dissections, 52
  was improved on, 31, 52
wall for 3D strip, 243
Wallace, William
  biography, 223
Wells, David
  four triangles to one, 47
Wheeler, Albert H.
  biography, 145
  pentagon to triangle, 141
  was improved on, 237
Wheeler, Roger F.
  biography, 250
  cubes for $3^3 + 4^3 + 5^3 = 6^3$, 248
Wotherspoon, George
  biography, 128
  double P-slide, 33
  was improved on, 127